TREATMENT OF PETROLEUM REFI
WASTEWATER WITH CONSTRUCTED WETLANDS

Hassana Ibrahim Mustapha

Thesis committee

Promotor
Prof. Dr P.N.L. Lens
Professor of Environmental Biotechnology
IHE Delft, Institute for Water Education, the Netherlands

Co-promotor
Dr J.J.A. van Bruggen
Senior Lecturer (Microbiology)
IHE Delft, Institute for Water Education, the Netherlands

Other members
Prof. Dr Karel J. Keesman, Wageningen University & Research
Prof. Dr Jos T.A.Verhoeven, Utrecht University
Dr Diederik P.L. Rousseau, Ghent University, Belgium
Dr Natalie Fonder, University of Namur, Belgium

This research was conducted under the auspices of the SENSE Research School for Socio-Economic and Natural Sciences of the Environment

Treatment of petroleum refinery wastewater with constructed wetlands

Thesis
submitted in fulfilment of the requirements of
the Academic Board of Wageningen University and
the Academic Board of the IHE Delft, Institute for Water Education
for the degree of doctor
to be defended in public
on Friday 29, June 2018 at 4 p.m.
in Delft, the Netherlands

by

Hassana Ibrahim Mustapha
Born in Lokoja, Nigeria

Published by:
CRC Press/Balkema
Schipholweg 107C, 2316 XC, Leiden, the Netherlands
Pub.NL@taylorandfrancis.com
www.crcpress.com – www.taylorandfrancis.com

ISBN: 978-1-138-32439-8
ISBN: 978-94-6343-844-5
DOI: https://doi.org/10.18174/444370

Dedication

To the two most precious gems in my life: Raliya Abubakar and Hauwa Kulu, I am what I am with your support and unconditional love.

To my husband, Mustapha, your unwavering support made this possible.

Acknowledgement

Firstly, I wish to express my gratitude to my promotor, Prof. Piet N. L. Lens for his guidance, patience, constructive criticisms, advice, unrelenting effort and in-depth supervision throughout the period of my PhD programme. I am truly honoured to have worked with him. I also thank my mentor, Dr. J. J. A van Bruggen for his thorough review, comments and critical supervision that has contributed greatly to the success of the programme. And my profound gratitude goes to Prof. Diederik Rousseau for providing insight into the research.

My appreciation goes to the management of KRPC, Kaduna (Nigeria) for providing the experimental site which contributed greatly to the outcome of the research and the staffs especially Dr. Ibrahim El-Idris and Hajia Aisha Bashir of the Water Quality Unit for their assistance.

And to the staffs of IHE DELFT including Jolanda (for logistics), the laboratory for their assistance, training and insight into the research, departmental secretaries (EEWT and WSE), members of Pollution Prevention and Resource Recovery core group (especially Dr. Eldon Raj for coordinating the scientific meetings), Anique (Beadle) and Paula, social cultural office (accommodation matters and other issues), library, ICT desk (computer issues) and finances. I am also grateful to my IHE family (Eman Sheikh, Zahra Musa, Khaphilat, Yetty, Pedi, AbdulHafiz, Ishaq Bello, Adebayo, Kemi and others not mentioned here) that I have really grown fond of, thank you for making my stay in the Netherlands worthwhile.

I specially acknowledge the assistance rendered by Dr. Johnson Ayayi (Mann-Kendall test) through Dr. Segun Adebayo, my dear brother Mikhail (editing and typesetting of the thesis) and my precious students now my colleagues: Felix (computer), Shaba (Math lab), AbdulHakeem and AbdulWahab (SPSS) as well as AbdulHakeem Ango, Beida, Abdullahi and Nafiz (fieldwork). I am greatly indebted to you all for contributing to the success of this research.

Also, to my numerous colleagues for their support, advice, assistance, quick responses to my many requests and demands, materials and encouragement throughout the period. I am truly indebted to you and may God reward you abundantly.

Likewise, my friends for being there through thin and thick; loving and understanding me in spite of my short comings. I extend my profound gratitude to your husbands for their relentless support and contributions.

And finally to my family; firstly, in the loving memory of my late father, Alhaji Muhammad Inuwa Ibrahim, jazakumullahu khayrah. My appreciation goes to my papa, the Ona of Abaji, for his enormous support, prayers, encouragement and firm believe in me. Also in loving memory of my Ba'aba, jazakumullahu khayrah. And to Hajia Hauwa kulu, I do not have the right words that can express my sincere gratitude but thank you mama dear for inevitably being there. To all my older and younger sisters and brothers; aunties and uncles; cousins; nephews; nieces and grandparents; I feel I am the luckiest person on earth to have you as family. I could not have done it without your immense efforts towards the success of this programme.

Furthermore, to my in-laws: father, mother, sisters, brothers and the entire extended family; sincerely this PhD programme would have not taken place without your permission, continuous support and contributions. Thank you so much. To my dear children, thank you for your patience, love, encouragement and inspiration.

And finally to my husband; how could I have achieved this without your unwavering support. There are no words that can describe my heartfelt appreciation, however, thanks for your physical, intellectual, inspirational, emotional and spiritual contributions. I love you.

All praises and gratitude is to Allah! The All-Knowing! The All-seeing and the All-Hearing!

Summary

The use of constructed wetlands (CWs) for polishing of petroleum refinery wastewater in Nigeria was evaluated. Secondary treated petroleum refinery wastewater from a refinery (Kaduna, Nigeria) was characterized with different types of organic and inorganic pollutants (Chapter 3). Vertical subsurface flow (VSSF) CWs planted with locally available macrophytes (*Cyperus alternifolius* and *Cynodon dactylon*) were designed and built for polishing of secondary treated refinery wastewater in terms of organic matter, nutrients and suspended solids removal (Chapter 4). The tertiary treated refinery wastewater did, however, not meet effluent discharged compliance limits in terms of total suspended solids (TSS), biochemical oxygen demand (BOD_5), chemical oxygen demand (COD) and ammonium-N (NH_4^+-N) removal.

Typha latifolia planted-VSSF CWs could, however, treat TSS, BOD_5, COD and NH_4^+-N in the petroleum refinery wastewater to below World Health Organization and Federal Environmental Protection Agency (Nigeria) effluent discharge limits of 30 mg/L for TSS, 10 mg/L for BOD_5, 40 mg/L for COD and 0.2 mg/L for NH_4^+-N (Chapter 5). *T. latifolia*-planted VSSF CW achieved higher removal efficiencies for all parameters measured in comparison to *C. alternifolius* and *C. dactylon* planted-VSSF CWs. In addition, the *T. latifolia*-planted VSSF CW had the best heavy metal removal performance, followed by the *C. alternifolius*-planted VSSF CW and then the *C. dactylon*-planted VSSF CW (Chapter 6). The accumulation of the heavy metals in the plants accounted for only a rather small fraction (0.09 - 16 %) of the overall heavy metal removal by the wetlands. Coupling a horizontal subsurface flow (HSSF) CW to the VSSF CW (hybrid CW) further improved effluent quality with an overall BOD_5 and PO_4^{3-}-P removal efficiency of, respectively, 94% and 78% (Chapter 5).

Diesel contaminated wastewater was treated in the hybrid CWs spiked with three different nutrient concentrations. Numerical experiments were performed to investigate the biodegradation of the diesel compounds in the synthetic contaminated wastewater by the duplex-CWs using constructed wetland 2D. The VF CWs had a higher removal efficiency than the HFF CWs and the hybrid CW showed higher removal efficiencies in the days with nutrient application than the days without nutrient application (Chapter 8).

This study showed that VSSF CWs planted with *T. latifolia*, *C. alternifolius* and *C. dactylon* can be used for the removal of suspended solids, organic contaminants and heavy metals from secondary refinery wastewater under tropical climate conditions. Especially *T. latifolia*-planted hybrid CWs are viable alternatives for the treatment of secondary refinery wastewater to below standards of the World Health Organization and Federal Environmental Protection Agency (Nigeria) under the prevailing climatic conditions in Nigeria.

Samenvatting

Het gebruik van constructed wetlands (CW's) voor het polijsten van afvalwater van olieraffinaderijen in Nigeria werd geëvalueerd. Secundair behandeld afvalwater van een raffinaderij (Kaduna, Nigeria) bevat verschillende soorten organische en anorganische verontreinigende stoffen (hoofdstuk 3). Verticale ondergrondse stroming constructed wetlands (VSSF CW's) beplant met lokaal beschikbare macrofyten (*Cyperus alternifolius* en *Cynodon dactylon*) werden ontworpen en gebouwd voor het polijsten van secundair behandeld raffinaderijafvalwater, met name verwijdering van organisch materiaal, nutriënten en zwevende stoffen (hoofdstuk 4). Het tertiaire gezuiverde afvalwater van de raffinaderij voldeed echter niet aan de lozingsnormen voor totale gesuspendeerde vaste stoffen (TSS), biochemische zuurstofvraag (BZV$_5$), chemische zuurstofvraag (CZV) en ammonium stikstof (NH$_4^+$-N).

Met *Typha latifolia* beplante VSSF CW's zouden echter TSS, BOD$_5$, COD en NH$_4^+$-N in het afvalwater van de aardolieraffinaderij kunnen zuiveren tot onder de lozingsnormen van de Wereldgezondheidsorganisatie en het Federale Milieubeschermings Agentschap (Nigeria) van 30 mg/L voor TSS, 10 mg/L voor BZV$_5$, 40 mg/L voor CZV en 0.2 mg/L voor NH$_4^+$-N (Hoofdstuk 5). *T. latifolia* beplante VSSF CW behaalden hogere verwijderingsefficiënties voor alle parameters in vergelijking met *C. alternifolius* en *C. dactylon* beplante VSSF CW. Bovendien hadden de met *T. latifolia* beplante VSSF CW de beste verwijderingsprestaties voor zware metalen, gevolgd door de *C. alternifolius* beplante VSSF CW en vervolgens de *C. dactylon* beplante VSSF CW (Hoofdstuk 6). De ophoping van de zware metalen in de planten vertegenwoordigde slechts een vrij kleine fractie (0.09 - 16%) van de totale verwijdering van de zware metalen door de wetlands. Het koppelen van een horizontale ondergrondse stroom constructed wetland (HSSF CW) met de VSSF CW (hybride CW) verbeterde de effluentkwaliteit met een totale BOD$_5$ en PO$_4^{3-}$-P verwijderefficiëntie van respectievelijk 94% en 78% (Hoofdstuk 5).

Diesel verontreinigd afvalwater werd behandeld in de hybride constructed wetlands met drie verschillende nutriëntenconcentraties. Numerieke experimenten werden uitgevoerd om de biodegradatie van de dieselverbindingen door de duplex-CW's te onderzoeken. De VF CW's hadden een hogere verwijderingsefficiëntie dan de HFF CW's en de hybride CW's toonden hogere verwijderingsefficiënties in de dagen waarop nutriënten werden gedoseerd (Hoofdstuk 8).

Deze studie toonde aan dat VSSF CWs beplant met *T. latifolia*, *C. alternifolius* en *C. dactylon* kunnen worden gebruikt voor het verwijderen van gesuspendeerde vaste stoffen, organische verontreinigingen en zware metalen uit secundair raffinaderijafvalwater onder tropische klimaatomstandigheden. Met name *T. latifolia* beplante hybride CW's zijn een volwaardig alternatief voor de behandeling van afvalwater van secundaire petroleum raffinage tot onder de normen van de Wereldgezondheidsorganisatie en het Federale Milieubeschermings Agentschap van Nigeria.

Table of Contents

Ch. 1. General Introduction

1.1. Background of the study

Nigeria is faced with rapid urbanization and industrialization which is producing huge amounts of industrial wastes with little or no regulations on their handling. These industrial wastes are disposed on arable lands and in rivers (Sikder et al., 2013; Saien, 2010; Agbenin et al., 2009). Waste contributes to a variety of toxic effects on living organisms in the food chain by bioaccumulation and biomagnification (Jadia and Fulekar, 2009). The accumulation of discharged of unwanted industrial by-products released into the environment without treatment is harmful (Agarry et al., 2008; Akpan et al., 2008). Effective treatment is needed before industrial wastewaters are discharged into the environment or reused for irrigation, livestock, groundwater recharge and for many other purposes (Spacil et al., 2011).

Oil exploration and exploitation activities have contributed to the economic growth of Nigeria but has also resulted in several incidences of oil spills causing increased environmental pollution and degradation (Eyo-Essien, 2008). It was estimated by Kadafa (2012) that 9 - 13 million barrels of oil have been spilled into the Niger Delta of Nigeria, the largest wetland in Africa, in the last 50 years and a total range of 0.7-1.7 million tons of petroleum have been discharged into the oceans, seas and rivers through anthropogenic activities. Oil spills can lead to displacement of air pore spaces in soil particles (Mustapha et al., 2015), causing wide deforestation and pollution of both water bodies and terrestrial ecosystems (Mmon and Deckor, 2010; Jadia and Fulekar, 2009; Li and Yang, 2008), eventually resulting into global environmental issues (Xia et al., 2003).

Large volumes of fresh water are extracted and consumed for use by petroleum industries in the process of refining crude oil and as cooling agent (Saien, 2010; Shpiner, et al., 2009; Allen, 2008). Similarly, large volumes of wastewater are generated (Mustapha et al., 2015). Diya'uddeen et al. (2011) reported that the produced refinery effluent during processing amounts to 0.4 - 1.6 times the amount of crude oil processed, and that this estimate is based on the yield of 13 million m^3 (84 million barrels per day) of crude oil and a total of 5.3 million m^3 of effluent is generated globally.

Scientists and engineers have been investigating the ability of plants as remediation alternatives for treating a wide variety of pollutants in contaminated waters (Mustapha et al., 2015; Marchand et al., 2010; Jadia and Fulekar, 2009; Imfeld et al., 2009). Natural treatment systems are environmentally friendly and use solar-driven biological processes to treat pollutants (Wenzel, 2009). They provide a less intrusive approach than the harsh conventional methods like incineration, thermal vaporization, solvent washing or other soil washing techniques, which can destroy the biological component of the soil or change the chemical or physical characteristics of the soil (Lin and Mendelssohn, 2009; Hinchman et al., 1996).

1.2 The Kaduna Refining and Petrochemical Company (KRPC), Kaduna

The Kaduna Refining and Petrochemical Company (KRPC) is located in Kaduna State (Nigeria). Kaduna state is situated in the Northern guinea savannah ecological zone of Nigeria. It lies between Latitude 9 ^{0} N and 12 ^{0} N and Longitude 6 ^{0} E and 9 ^{0} E of the prime meridian. The climatic condition is categorized by constant dry and wet seasons. The rains begin in April/May and ends in October, while the dry season starts late October and stops in March of the subsequent year. The mean annual rainfall is between 1450 - 2000 mm with a mean daily temperature regime ranging from 23 to 25^0 C, and a relative humidity varying between 20 and 40% in January and 60 and 80% in July. It has a solar radiation ranging between 20.0 - 25.0 Wm^{-2} day^{-1} (Mustapha et al., 2015; Emaikwu et al., 2011).

KRPC occupies a land area of 2.89 square kilometers approximately 15 km Southeast of Kaduna city with an elevation of approximately 615 m above mean sea level (Bako et al., 2008). It was commissioned in 1980, with an initial capacity of 100,000 Barrels per Stream Day (BPSD) as the third refinery in Nigeria in order to cope with the tremendous and growing demand for petroleum products (Bako et al., 2008). It has a Fuel Plant Crude Distillation Unit I (CDU I) and a Lube Plant Crude Distillation Unit II (CDU II) (Jibril et al., 2012). The Fuel Plant was designed to process 50,000 BPSD, it was later increased by an additional 60,000 BPSD bringing the total refinery installed capacity to 110,000 BPSD (Jibril et al., 2012; Bako et al., 2008). It processes Escravos light crude oil and Ughelli Quality Control Centre (UQCC) crude oil while the Lube plant has capacity to process 50,000 BPSD of imported paraffin rich crude oil for the manufacture of lubricating oils. The crude oils available in Nigeria cannot produce the whole range of fractions for the production of lubricating oils, hence Nigerian National Petroleum Corporation (NNPC) imports from Venezuela, Kuwait or Saudi Arabia (Jibril et al., 2012).

Figure 1.1: Google map of Kaduna refining and petrochemical company (KRPC), Kaduna. Note: S – Experimental site and R- Romi River.

1.3 Problem statement

Contamination of water and lands by petroleum chemicals is a global issue and is of particular concern in developing countries such as Nigeria where industrial pollution is one of its major problems. In addition, the release of untreated or partially treated petroleum refinery effluents into the environment leads to pollution of the terrestrial and aquatic ecosystems, an unaesthetic environment, high cost of wastewater treatment, loss of farmlands, fishes, portable water and means of the livelihood particularly in the oil polluted-affected community.

The polluted sites are fast increasing and need to be addressed by low cost and effective technology such as constructed wetlands (CWs) (Schröder et al., 2007). CW technology is ecologically friendly and a more economical and easier way of treating wastewater. However, the use of CWs to treat hydrocarbon contaminants in refinery wastewater and other wastewater types in Nigeria is relatively new as compared to conventional treatment systems. The use of CWs has also been largely ignored by developing countries.

CWs have been used for decades for the treatment of different types of wastewater and have been recognized as a reliable wastewater treatment technology. They represent a suitable solution for treatment of many types of wastewater (Vymazal, 2011). There is a need, therefore, to use CWs to treat petroleum-contaminated wastewater in developing nations such as Nigeria where good water quality and resources are scarce and effective low cost wastewater treatment strategies are critically needed. Existing technologies do not sufficiently address the increasing pollution situation, also some chemical approaches such as advanced oxidation steps cannot be adopted by most wastewater treatment plants (Schroder et al, 2007).

1.4. Research objectives of the thesis

1.4.1. General objectives

Typha latifolia and *Phragmites australis* is commonly used for petroleum wastewater treatment. Few works on treatment of petroleum contaminated wastewater have been reported on *Cyperus alternifolius* and none on *Cynodon dactylon* (L.) Pers. to the best of the author's knowledge. This study, therefore, seeks to examine the suitability of *T. latifolia, C. alternifolius, C. dactylon* and *P. australis,* for the treatment of petroleum refinery wastewater. Thus, the overall aim of this study is the polishing of secondary refinery effluent to below compliance limits recommended by regulating authorities before discharge into environment or for reuse purposes.

1.4.2. Specific objectives

In order to address the problems above, the following specific objectives of this research work are to:

1. Characterize the effluent from the Kaduna refining and petrochemical company (KRPC) at the point of discharged into the Romi River (Chapter 3);
2. Design, construction, operation and monitoring of planted and unplanted VSSF, HSSF and hybrid CWs (Chapters 4 and 5);
3. Examine the effectiveness of *T. latifolia, C. alternifolius* and *C. dactylon* for the treatment of petroleum-contaminated wastewater with respect to nutrient (ammonium, nitrate and phosphate), organic pollutants (BOD, COD, TPH, phenol, oil & grease) and heavy metals (Cr, Cd, Pb, Cu, Zn and Fe) (Chapters 4, 5, 6 and 7);
4. Examine the colonization characteristics of the microorganisms' active in the rhizosphere (Chapter 4);

5. Determination of bioaccumulation and translocation factors of *T. latifolia, C. alternifolius* and *C. dactylon* (Chapters 6 and 7) and

6. Use duplex CWs planted with *P. australis* to treat petroleum contaminants (BTEX compounds) (Chapter 8).

1.5.   Research questions

In designing, constructing, and operating the experimental wetlands to solve the identified problems, this study seeks to answer specific questions that may arise such as:

- Can constructed wetlands effectively and efficiently treat petroleum-contaminated wastewater to compliance limit?
- Are emergent macrophytes (*Typha latifolia, Phragmites australis, Cyperus alternifolius* and *Cynodon dactylon*) suitable for the treatment of petroleum-contaminated wastewater?
- What are the roles of the different components of constructed wetlands in removal of the pollutants of interest?
- Which are the most important processes associated with pollutant removal in constructed wetlands?
- What are the possibilities of modeling and simulating the contaminant removal potentials of this natural treatment system?
- What is the possibility of acceptance of this technology by petroleum and petrochemical related industries particularly in developing countries?

In order to achieve the specific objectives of this research, and find answers to the research questions raised, four experiments were designed consisting of five experimental phases. These phases are defined based on the specific objectives.

1.6.   Significance of the study

The treatment of petroleum contaminated water with CWs using a vertical and horizontal subsurface flow system is the main focus of this research work. Therefore, the significance of this study is:

1. The study area is in Nigeria. Nigeria is a major oil-producing country. Water and soil pollution from petroleum processing is a persistent problem in Nigeria. The use of CWs for treating petroleum-contaminated wastewater is an ecologically friendly alternative to conventional technologies. This treatment system is especially ideal for developing nations such as Nigeria where quality water is scarce.

2. The use of CWs for wastewater treatment is an emerging technology in Nigeria, there is no recorded work to the best of the author's knowledge on the use of this natural technology for the treatment of petroleum-contaminated wastewater in Nigeria. This study is an unprecedented research for this type of wastewater in Nigeria. The study will bring about awareness of the use of constructed wetland treatment system to Nigeria and contribute to the existing knowledge.

3. CW technology is simple, cost effective, can easily be maintained when compared with conventional methods. The technology can be adopted for use by oil-producing industry in Nigeria and other developing nations where effective, low cost wastewater treatment strategies are critically needed;

4. The research will assist in reducing the level of hazardous constituents into water bodies as well as soil and assure improved wastewater quality by the discharge of treated wastewater into the environment;

5. The adequately treated wastewater from CW systems can be reused and/or safely discharged into water bodies, this can drastically reduce the cost of production of potable water;

6. Fewer polluted water bodies will ensure an environment that is conducive for fishes and all forms of aquatic life as well as improve the livelihood of the polluted-affected community;

7. Health problems and diseases associated with the discharge of untreated or inadequately treated wastewater will be minimized.

1.7. Outline of the thesis

This thesis has been divided into nine chapters.

Chapter 1 gives a background to the dissertation including the problems associated with use of conventional techniques for refinery wastewater treatment. It also gives the objectives to be addressed, research questions as well as the description of the study site and outline of the thesis.

Chapter 2 reviews the relevant literature on the use of CWs and conventional treatment systems for refinery wastewater treatment. In addition, it gives an overview of the roles of the components of CWs and types of CWs. Furthermore, it put emphasis on the benefits of CWs for developing countries.

Chapter 3 presents the characterization of effluent from Kaduna refining and petrochemical company (KRPC), Kaduna (Nigeria). The generated data give baseline information on the quality of effluent discharged by KRPC. The results show that conventional treatment methods cannot properly remove all the contaminants of concern in their treated refinery effluents.

Chapter 4 describes the design, experimental setup, plant establishment, operation and monitoring of vertical subsurface flow constructed wetlands (VSSF CWs). Data from influent and effluent parameters from two locally available macrophytes (*Cyperus alternifolius* and *Cynodon dactylon)* planted VSSF CWs in terms of temperature, pH, COD, BOD, nutrients, TDS are presented in this chapter.

Chapter 5 presents the optimization of the performance of *Cyperus alternifolius* and *Cynodon dactylon* treatment wetlands by introducing a third plant *Typha latifolia* in combination with a horizontal subsurface flow constructed wetlands (HSSF CWs). This chapter present an improved performance combined with the existing data.

Chapter 6 describes the fate of heavy metals in CWs in phytoremediation of heavy metals from refinery wastewater stream. Bioaccumulation and translocation factors of *T. latifolia, C. alternifolius* and *C. dactylon* are presented here.

Chapter 7 presents treatment of petroleum organic pollutants (TPH, phenol and oil and grease) by VSSF CWs. Based on the results from chapter 5, *T. latifolia* was chosen for this experiment. Bioaccumulation and translocation of the contaminants of concern in plant parts were determined.

Chapter 8 describes the treatment of simulated refinery wastewater with duplex constructed wetlands. Effect of the addition of nutrients on the performance of duplex-CWs for the removal of petroleum compounds in simulated refinery effluent were considered in this chapter.

And finally, Chapter 9 integrates the outcomes of the thesis, compares and discusses the results from the different chapters as well as provides some general conclusions.

1.8. References

Agarry, S. E., Durojaiye, A. O., Yusuf, R. O., Solomon, B. O., Majeed, O. (2008). Biodegradation of phenol in refinery wastewater by pure cultures of *Pseudomonas aeruginosa* NCIB 950 and *Pseudomonas fluorescence* NCIB 3756. International Journal of Environmental Pollution 32(1): 3-11. doi:10.1504/IJEP.2008.016894

Agbenin, J. O., Danko, M., Welp, G. (2009). Soil and vegetable compostional relationships of eight potentially toxic metals in urban garden fields from northern Nigeria. Journal of Science, Food and Agriculture 89(1): 49-54. doi:10.1002/jsfa.3409

Akpan, U. G., Afolabi, E., Okemini, K. (2008). Modelling and simulation of the effects of Kaduna refinery and petrochemical company on River Kaduna. AU Journal of Technology 12(2): 98-106.

Allen, W. E. (2008). Process water treatment in Canada's oil sands industry: II. A review of emerging technologies. Journal of Environmental Engineering and Science 7:499-524. doi:10.1139/S08-020

Bako, S. P., Chukwunonso, D., Adamu, A. K. (2008). Bioremediation of refinery effluents by strains of *Pseudomonas aerugenosa* and *Penicillium Janthinellum*. Journal of Applied Ecology and Environmental Research 6(3): 49-60. Retrieved from http://www.ecology.uni-corvinus.hu.

Diya'uddeen, B. H., Wan Daud, M. A., Abdul Aziz, A. R. (2011). Treatment technologies for petroleum refinery effluents: A review. Journal of Process Safety and Environmental Protection 89: 95-105. doi:10.1016/j.psep.2010.11.003

Emaikwu, K. K., Chiwendu, D. O., Sani, A. S. (2011). Determinants of flock size in broiler production in Kaduna State of Nigeria. Journal of Agricultural Extension and Rural Development 3(11): 202-211. ISSN- 2141 -2154. http:// academicjournals.org/JAERD

Eyo-Essien, L. P. (2008). Oil spill management in Nigeria: Changes of pipeline vandalism in the Niger Delta Region of Nigeria. Abuja-Nigeria: National Oil Spill Detection & Response Agency (NOSDRA). Retrieved from http://ipec.utulsa.edu/Conf2008/Manuscripts%20%20received/Eyo_Essien 2.pdf

Hinchman, R. R., Negri, M. C., Gatliff, E. G. (1996). Phytoremediation: using green plants to clean up contaminated soil, groundwater. International Tropical Meeting on Nuclear and Hazardous Waste Management. Spectrum 96, Seattle, WA: American Nuclear Society.

Imfeld, G., Braeckevelt, M., Kuschk, P., Richnow, H. H. (2009). Monitoring and assessing processes of organic chemicals removal in constructed wetlands. Chemosphere 74(3): 349-362. doi:10.1016/j.chemosphere.2008.09.062

Jadia, C., Fulekar, M. (2009). Phytoremediation of heavy metals: Recent techniques. African Journal of Biotechnology 8(6): 921-928.

Jibril, M., Aloko, D. F., Manasseh, A. (2012). Simulation of Kaduna refining and petrochemical company (KRPC) crude distillation unit (CDU I) using Hysys. International Journal of Advanced Scientific Research and Technology 1(2): 1-6.

Kadafa, A. A. (2012). Oil exploration and spillage in the Niger Delta of Nigeia. Civil Environmental Research 2(3): 38-51. Retrieved from www.iiste.org

Li, M. S., Yang, S. X. (2008). Heavy metal contamination in soils and phytoaccumulation in a manganese mine wasteland South China. Air, Soil and Water Research 1:31-41.

Lin, Q., Mendelssohn, I. A. (2009). Potential of restoration and phytoremediation with *Juncus roemerianus* for diesel-contaminated coastal wetlands. Ecological Engineering, 35(1), 85-91. Retrieved from http://dx.doi.org/10.1016/j.ecoleng.2008.09.010

Marchand, L., Mench, M., Jacob, D. L., Otte, M. L. (2010). Metal and metalloid removal in constructed wetlands, with emphasis on the importance of plants and standardized measurements: A review. Environmental Pollution, 158(12): 3447-3461. doi:10.1016/j.envpol.2010.08.018

Mmon, P., Deckor, T. (2010). Assessing the Effectiveness of land farming in the remediation of hydrocarbon polluted soils in the Niger Delta, Nigeria. Research Journal of Applied Sciences, Engineering and Technology 2(7): 654-660.

Mustapha, H. I., van Bruggen, J. J. A., Lens, P. N. L. (2015). Vertical subsurface flow constructed wetlands for polishing secondary Kaduna refinery wastewater in Nigeria. Ecological Engineering 84: 588-595. doi:10.1016/j.ecoleng.2015.09.060

NIMET. (2010). Nigeria Meteorological Agency, Nigeria. Kaduna, Kaduna, Nigeria.

Saien, J. (2010). USA Patent No. Pub. No.: US 20100200515 A1. Retrieved February 2011

Schröder, P., Navarro-Aviñó, J., Azaizeh, H., Azaizeh, H., Goldhirsh, A. G., DiGregorio, S., . . . Wissing, F. (2007). Using phytoremediation technologies to upgrade waste water treatment in Europe. Environmental Science and Pollution Research - International 14(7): 490-497. doi:10.1065/espr2006.12.373

Shpiner, R., Vathi, S., Stuckey, D. C. (2009a). Treatment of "produced water" by waste stabilization ponds: removal of heavy metals. Water Research 43: 4258-4268. doi:10.1016/j.watres.2009.06.004

Sikder, M. T., Kihara, Y., Yasuda, M., Mihara, Y., Tanaka, S., Odgerel, D., Mijiddorj, B., Syawal, S. M., Hosokawa, T., Saito, T., Kurasaki, M. (2013). River water pollution in developed and developing countries: Judge and assessment of physicochemical and characteristics and selected dissolved metal concentration. CLEAN-Soil Air Water 41(1): 60-68. doi:10.1002/clen.201100320

Spacil, M., Rodgers, J., Castle, J., Murray-Gulde, C., Myers, J. E. (2011). Treatment of Selenium in simulated refinery effluent using a pilot-scale constructed wetland treatment system. Water, Air, and Soil Pollution 221(1): 301-312. doi:10.1007/s11270-011-0791-z

Vymazal, J. (2011). Constructed wetlands for wastewater treatment: five decades of experience. Journal of Environmental Science and Technology 45(1): 61-69. doi:10.1021/es101403q

Wenzel, W. W. (2009). Rhizosphere processes and management in plant-assisted bioremediation (phytoremediation) of soils. Plant and Soil 321(1-2): 385-408. doi:10.1007/s11104-008-9686-1

Xia, P., Ke, H., Deng, Z., Tang, P. (2003). Effectiveness of constructed wetlands for oil-refined wastewater purification. In P. N. Truong & X. Hanping (Ed.), the 3rd International Conference on Vetiver and Exhibition : vetiver and water : an eco-technology for water quality improvement, land stabilization, and environmental enhancement. Vol. 22, pp. 649-654. Guangzhou: Guangdong Academy of Agricultural Sciences China Agriculture Press. Retrieved 2011, from http://trove.nla.gov.au/work/37498028?versionId=48942993

Ch. 2. Constructed wetlands to treat wastewater generated in conventional petroleum refining industry: a review

This chapter has been submitted to a Journal for publication entitled:
Hassana Ibrahim Mustapha., P. N. L. Lens (2018), "Constructed wetland to treat wastewater generated in conventional petroleum refining industry: A review"

Abstract

Petroleum refining industries extract large volumes of freshwater for the process of refining crude oil. Thus, large volumes of wastewater are generated. These wastewaters contain contaminants that are neurotoxic, carcinogenic and mutagenic. Petroleum related industries have to comply with strict regulations by regulatory authorities, these have led them to explore many treatment technologies for effluent management. However, conventional wastewater treatment technologies have higher energy and capital requirement than constructed wetland treatment systems. In addition, their processes can lead to incomplete decomposition of contaminants which can generate by-products that are often toxic to both humans and the environment. Thus, these by-products require further treatment. This paper reviews literature specifically on the use of constructed wetlands (CW) systems for treatment of petroleum contaminated wastewater from conventional oil and gas industry. CWs can effectively treat multiple contaminants in secondary and tertiary wastewater with less production of sludge. They also have the advantages of providing an ecologically friendly approach with low energy demand and operational cost. Therefore, they are a viable alternative to conventional wastewater treatment systems. Plants, microorganisms, substrate media and different flow types play important roles in the removal of pollutants from the refinery wastewater in CWs, however, microorganisms play the most crucial role in the transformation and mineralization of nutrients and organic pollutants. Many studies reported significantly higher removal efficiencies of petroleum pollutants in planted compared to unplanted gravel beds, and that subsurface flow CWs can achieve more biological treatment than surface flow CWs due to the higher specific surface area of the gravel bed. CWs can be intensified with forced aeration to enhance aerobic biodegradation rates of petroleum hydrocarbons in contaminated wastewaters to non-detectable levels.

Keywords: Petroleum, Wastewater, Contaminants, Constructed wetlands, Plants, Microorganisms, Substrates

Abbreviations and notations

Abbreviations

API	American Petroleum Institute
BOD_5	Biochemical Oxygen Demand at 20 °C over 5 days
BTEX	Benzene, Toluene, Ethyl-benzene, Xylene
COD	Chemical Oxygen Demand
$CBOD_5$	Carbonaceous biological oxygen demand
COCs	contaminants of concerns
CWs	Constructed wetlands
FWS	Free water surface flow
GRO	Gasoline range organics
HLR	Hydraulic Loading Rate
HRT	Hydraulic Retention Time
MTBE	Methyl tert-buthyl ether
NH_3	Ammonia
O&M	Operation and maintenance
PAHs	Poly aromatic hydrocarbons
SF	Surface flow
SSF	Subsurface flow
TDS	Total Dissolved Solids
TPH	Total petroleum hydrocarbon
TPH(DRO)	Total petroleum hydrocarbon (diesel range organics)
TSS	Total suspended solids
TKN	Total Kjeldahl Nitrogen

2.1 Introduction

Petroleum industry

Petroleum is made up of crude oil and natural gas. Crude oil is a complex mixture of hydrocarbons (Merkl et al. 2006; Abu and Dike 2008). Petroleum refining industry converts crude oil into finished products (Gousmi et al., 2016) such as transportation fuels (gasoline, diesel fuel, jet fuel, compressed natural gas and propane) (Gousmi et al. 2016), heating fuels (propane, liquefied petroleum gas, kerosene, heating oil, and natural gas), sources of electricity (natural gas and residual fuel oil) and petrochemicals (feedstocks for plastics, clothing and building materials) (Hamza et al. 2012). Refined products from crude oil comprise a varied range of compounds such as hydrocarbons, heavy metals, dye additives, antioxidants and corrosion inhibitors (Adewuyi and Olowu 2012). Consequently, refined products can show higher toxicity compared to crude oil due to alteration of metal speciation and the metals added to the matrix during the refining processes (Adewuyi and Olowu 2012). Based on their differential solubility in organic solvents (Abu and Dike 2008), there are four major petroleum compounds (Eke and Scholz 2008), namely saturated hydrocarbons (which are the primary components), aromatic hydrocarbons, resins (pyridines, quinolones, carbazoles, sulfoxides and amides) and asphaltenes (phenols, fatty acids, ketones, esters and porphyrins) (Das and Chandran 2011; Eke and Scholz 2008). Hydrocarbons originate from natural and man-made sources (Achile and Yillian 2010). They are potentially toxic chemicals, highly soluble and neurotoxic (Achile and Yillian 2010; Eke and Scholz 2008; Ogunfowokan et al. 2003). Due to their toxicity, aromatic hydrocarbons are of serious concern (Eke and Scholz 2008; Ogunfowokan et al. 2003), they are highly soluble and have the ability to readily migrate into groundwater (Eke and Scholz 2008) and the environment (Grove and Stein 2005). Moreover, these contaminants are not easily degraded by conventional treatment (Saien and Shahrezaei 2012).

2.1.1 Petroleum refining wastewater types

Petroleum refining industries extract large volumes of freshwater in the process of refining crude oil and as cooling agent (Nacheva 2011; Saien 2010; Shpiner et al. 2009a; Allen 2008). Consequently, large volumes of wastewater are generated (Nwanyanwu and Abu 2010; Diya'uddeen et al. 2011; Hamza et al. 2012; Diya'uddeen et al. 2012; Yu et al. 2013) as a result of production, storage, distribution and processing of petroleum, or by accidents due to spills from water/fuel mixtures, oil well drilling, leaks from underground storage or water collected from secondary containment and sumps (Saien 2010; Diya'uddeen et al. 2011; Jeon et al. 2011; Ballesteros Jr. et al. 2016).

The characteristics of petroleum wastewater from different refineries vary from one region of the globe to another depending on the region in which the crude oil was drilled, type of crude oil, its chemical composition, the different processes and the employed treatment mechanism (Valderrama et al. 2002; Wake 2005; Nwanyanwu and Abu 2010; Saien 2010; Nacheva 2011). This has significant impact on the character and quantity of contaminants entering a given refinery wastewater treatment system.

High rates of consumption of petroleum and its refined products will continue to generate effluents from petroleum refining processes, which, when discharged into water bodies would result in environmental pollution (Diya'uddeen et al. 2011; Zhao and Li 2011). The effects of the discharge of effluents include eutrophication, accumulation of toxic compounds in biomass and sediments, loss of dissolved oxygen in water (Paul 2011), contamination of drinking water and groundwater resources, thus endangering aquatic resources and human health as well as destruction of the natural landscape (Yu et al. 2013).

The types of wastewater stream generated by the petroleum refining industry can be classified into desalter water, cooling tower, spent catalyst, spent caustic, water used for flushing during maintenance and shut down, sour water and other residuals (Al Zarooni and Elshorbagy 2006; Nacheva 2011; Adewuyi and Olowu 2012; EPA n.d.) from desalting, catalytic cracking, astripping steam, sanitary, lube oil and asphalt (EPA n.d; Al Zarooni and Elshorbagy 2006). These wastes pose major problems that are a challenge faced by the petroleum industry, imposing the need to recover oil as well as to prevent the discharge of oily wastewater into the environment (Veil et al. 2004). Table 2.1 presents the types of wastewater generated in refining petroleum and their characteristics. These include refinery wastewater, brackish oilfield produced water, heavy-oil produced water, petroleum hydrocarbon contaminated water, sour water, produced water and diesel contaminated wastewater.

Produced water is the largest volume of wastewater generated by the petroleum industry (Veil et al. 2004; Asatekin and Mayes 2009). It is described as water from an oil well after its separation from oil in American Petroleum Institute (API) separators (Murray-Gulde et al. 2003; Shpiner et al. 2009a). Refinery wastewater is the wastewater generated from refining crude oil and manufacturing fuels, lubricants and petroleum intermediates (Mustapha et al.

2015). Other types of petroleum-contaminated wastewaters may include ballast water from ships, storm water and runoff from roads (Veenstra et al. 1998).

2.1.2 Petroleum contaminants

Petroleum-contaminated wastewaters contain different types of organic and inorganic pollutants with varying levels of contamination (Table 2.1). Petroleum wastewater is characterized by a range of pollutants (Mustapha et al. 2015) including organics, such as dispersed oil (Shpiner et al. 2009b; Nacheva 2011), oil and grease (Nwanyanwu and Abu 2010; Diya'uddeen et al. 2011) and heavy oil (viscosity > 100 mPas) (Ji et al. 2007), polycyclic aromatic hydrocarbons (PAH) (Fountoulakis et al. 2009; Uzoekwe and Oghosanine 2011; Hamza et al. 2012; Tromp et al. 2012), phenols (Agarry et al. 2008; Otokunefor and Obiukwu 2010; Hamza et al. 2012), and inorganics such as ammonia (NH_3) (Wallace 2001; Mustapha et al. 2015) and heavy metals (Ali et al. 2005; Jadia and Fulekar 2009; Mustapha et al. 2011; Mustapha et al. 2018). Some crude oils contain small quantities of metals that may require special equipment for refining the crude. In addition, oil and gas may contain sulphur and carbon dioxide that needs to be removed before marketing (Uzoekwe and Oghosanine 2011).

Wastes containing petroleum compounds, nutrients and other toxic compounds should be properly treated prior to discharge into the receiving water bodies (Abdelwahab et al. 2009) because these substances may pose serious hazards to the environment (Diya'uddeen et al. 2011) as well as their immediate damages to the organisms (Brito et al. 2009). The toxicity of petroleum refinery effluent has been reported in many studies (Abdelwahab et al. 2009; Das and Chandran 2011; Diya'uddeen et al. 2011). However, the toxicity depends on a number of factors, including quantity, volume and variability of discharge (Nwanyanwu and Abu 2010).

2.1.2.1 Organic pollutants

The discharge of wastewater with a high organic matter content into the aquatic environment results in the depletion of oxygen (Diya'uddeen et al. 2011). Organic pollutants produced by industrial activity such as BTEX (benzene, toluene, ethylbenzene and xylene) (Wallace 2001; Mazzeo et al. 2010), PAH and linear alkylbenzene sulfonates (Fountoulakis et al. 2009), chlorinated hydrocarbons (Haberl et al. 2003), benzene and methyl tert-buthyl ether (MTBE) (De Biase et al. 2011) have been successfully removed or retained by CW systems (Table 2.1). For example, Mustapha et al. (2015) analyzed secondary refinery wastewater which contained 106 (± 58.9) mg/L biological oxygen demand (BOD) and 232 (± 121.2) mg/L chemical oxygen demand (COD) in the pretreated influent. Czudar et al. (2011) also analysed petrochemical wastewater with measured BOD and COD in the influent ranging between 42 to 131 mg/L in spring, 42 to 144 mg/L in summer and 32 to 101 mg/L in autumn. The refinery wastewater characterized by Aslam et al. (2007) had concentrations of 109 - 197 mg/L BOD, 200 - 258 mg/L COD and 6 mg/L phenol.

There are fewer publications on the treatment of organic pollutants in petroleum refining wastewater with CWs compared with other types of wastewater, i.e. domestic, tannery, textile, abattoir, food processing, agricultural. BOD and COD concentrations are indicators of the level of organic compounds in a wastewater (Nwanyanwu and Abu 2010). Excessive level of BOD/COD in wastewater released into water bodies will reduce the level of dissolved oxygen, and low levels of dissolved oxygen can induce fish kills and reduce reproduction rates in aquatic life (Biswas, 2013). Also, the presence of high concentrations of organic pollutants in waste streams are toxic to plants and microorganisms (Knight et al., 1999). De Biase et al. (2011) investigated volatile organic compounds in contaminated groundwater next to refineries and chemical plants using vertical flow filters and vertical flow CWs. Their influents contained 20 ($\pm$ 2) mg/L of benzene and 39.0 ($\pm$ 0.5) mg/L of MTBE.

Petroleum refinery wastewater contains aliphatic and aromatic petroleum hydrocarbons at different concentrations (Saien and Shahrezaei 2012). The analysis of pre-treated refinery wastewater samples collected at the inlet of the biological treatment unit in the Kermanshah (Iran) refinery plant by Saien and Shahrezaei (2012) detected COD in the range of 200-220 mg/L, which consisted of methyl-tetrabutyl ether, phenol, 2,3,5,6-tetramethylphenol, naphthalene, xylene, tetradecane, 4-chloro-3-methylphenol and 3-*tert*-butylphenol.

Oil and grease (O&G) clog drain pipes and sewer liners, causing unpleasant odours and also corrode sewer lines under anaerobic conditions (Diya'uddeen et al. 2011). In addition, O&G in wastewater can cause depletion of dissolved oxygen and loss of biodiversity in the receiving water bodies (Mohammed et al. 2013). Polycyclic aromatic hydrocarbons are highly toxic, carcinogenic and mutagenic to microorganisms, organisms and humans (Fountoulakis et al. 2009; Zheng et al. 2013). Phenolic compounds are a serious problem due to their poor biodegradability and high toxicity (Abdelwahab et al. 2009). These compounds are harmful to organisms at concentrations as low as 0.05 mg/L (Nwanyanwu and Abu 2010) and carcinogenic to humans (Abdelwahab et al. 2009; El-Ashtoukhy et al. 2013; Zheng et al. 2013). In addition, phenol can reduce the growth and the reproductive capacity of aquatic organisms (Zheng et al. 2013).

Table 2.1: Characteristics of produced petroleum refinery wastewaters (mg/L, all except pH)

Types of wastewater	Influent concentrations													Reference
	COD	BOD	O&G	Benzene	Toluene	Other organics	BTEX	MTBE	P	N	TDS	Fe	pH	
Refinery wastewater	232 (± 121.2)	106 (± 58.9)	-	-	-	-	-	-	4.0 (± 2.0)	1.81 (± 1.6)	255.5 (± 70.3)	-	7.3 (± 1.6)	Mustapha et al. 2015
	79 - 130	34 - 93	-	13 - 32	-	-	1.2 - 5.9	1.2 - 5.9	0.1 - 2.7	37 - 63	-	-	-	Seeger et al. 2011;
	165 - 347	109 - 197	24 - 66	-	-	-	-	-	-	-	1167 - 2850	4 - 7.5	8.3 - 10.5	Aslam et al. 2007
	-	-	-	13.96 (±20)	-	-	2.97 (±0.82)	-	1.20 (± 0.75)	51 (± 9.34)	-	6.73 (±2.36)	7.45 (±0.35)	van Afferden et al. 2011
	-	-	-	0.17	-	-	0.47	-	-	-	-	33.3	-	Wallace et al. 2011
	200-220	-	-	-	-	-	-	-	-	-	-	-	-	Saien and Shahrezae, 2012
	-	-	-	-	169	40	1,123	-	-	-	-	-	-	Mazzeo et al. 2010
Oilfield produced water	1050 - 1350	-	-	-	-	-	-	-	-	-	-	-	-	Shpiner et al. 2009b
Heavy-oil produced water	390 (± 124)	32 (±8.5)	20 (± 4.6)	-	-	-	-	-	0.07 (± 0.03)	-	-	-	7.88 (± 0.40)	Ji et al. 2007
Oil-refined wastewater	132 - 196	26 - 78	36 - 61	8.97 – 9.30	-	-	-	-	-	22 - 29	-	-	7.10 – 7.21	Xia et al. 2003
Oilfield produced water	-	14693	1213	-	-	-	12.393	-	-	-	43048	-	-	Davis et al. 2009

Sample														Reference
Refinery diesel-range organics	-	3 - 64	-	0.001 - 2.090	<,0.001 -1.519	<,0.001 - 0.522	0.127 - 6.680	1.12 - 1.38	-	-	2116 - 2818	4.3 - 22.0	-	Bedessem et al. 2007
Petroleum Storage Tank	-	16000	-	-	-	-	0.00047	-	-	230	-	40	-	Wallace and Davis 2009
Petroleum-contaminated water	-	-	-	10.2 ± 3.8	0.002 ± 0.001	0.019 ± 0.017 1	0.88 ± 0.32	0.009 ± 0.004	1.80 (± 0.74)	27.1 (± 8.0)	-	3.14 (± 0.71)	-	Stefanakis et al. 2016
Leaking underground petroleum wastewater	-	-	-	66	-	-	-	-	-	-	-	-	-	Ballesteros Jr. et al. 2016
Refinery and chemical wastewater	-	-	-	20 (± 2)	-	-	39 (± 0.5)	-	-	-	-	-	-	De Biase et al. 2011
Petroleum hydrocarbon contaminated water	-	-	-	0.3	-	-	-	-	-	-	-	40	8.3	Wallace and Davis 2009
Hydrocarbon - contaminated wastewater	-	-	-	1300	-	-	-	-	-	-	-	-	-	Eze and Scholz 2008
Petroleum refinery wastewater	-	-	-	-	169	40	1,123	-	-	-	-	-	-	Mazzeo et al. 2010

2.1.2.2 Heavy metals

Metal contamination is a major environmental problem, especially in the aquatic environment (Chorom et al. 2012; Ho et al. 2012; Papaevangelou et al. 2017). Unlike organic pollutants, metals in wastewater are not degraded through biological processes (Yang et al. 2006). Many heavy metals are toxic both in elemental and soluble form (Jadia and Fulekar 2009). The most toxic metals are Cd, Pb, Hg, Ag and As (Hashim et al. 2011). Several authors have reported the presence of heavy metals in petroleum-contaminated wastewater (Gillespie Jr. et al. 2000; Moneke and Nwangwu 2011; Mustapha et al. 2017) as well as their hazardous effects (Calheiros et al. 2008; Hashim et al. 2011; Qasaimeh et al. 2015).

Cr, Cd and Pb are non-essential elements to plants and cause toxicity at multiple levels (Calheiros et al. 2008; Mustapha et al. 2018). Cu and Zn are essential elements for organisms (Song et al. 2011), however, these two metals become poisonous at excessive concentrations of 1 - 2 mg Cu /L and 3 - 5 mg Zn/L in waterbodies (Korsah 2011; Mebrahtu and Zerabruk 2011; Kumar and Puri 2012). In addition, iron (Fe) is important for all forms of life (Jayaweera et al. 2008). However, excessive doses of Fe > 1.6 mg/L can lead to hemorrhagic and sloughing of mucosa areas in the stomach of humans (Jayaweera et al. 2008). Ni is an essential trace element for plant growth and also a known human carcinogen at excessive levels (Chorom et al. 2012). Cr (VI) exhibits high toxicity, mobility, water solubility (Papaevangelou et al. 2017) and is carcinogenic (Rezaee et al. 2011). Wastewater that contains metals should be treated prior to discharge into the environment (Gillespie Jr. et al. 2000; Cheng et al. 2002; Rezaee et al. 2011).

Plants have the natural ability to uptake metals. The removal of heavy metals from CWs is by microbiota uptake (Khan et al. 2009), plant uptake as well as adsorption onto media and sediments in the system (Qasaimeh et al. 2015). Other processes that can contribute to heavy metal removal from wastewater in CWs are biosorption, bioaccumulation, redox transformation, dissimilatory sulphate reduction (Šíma et al. 2015) and precipitation as insoluble salts (Cheng et al. 2002).

2.1.2.3 Nutrients

Several authors such as Huddleston et al. (2000); Moreno et al. (2002); Nwanyanwu and Abu (2010); Moneke and Nwangwu (2011); Uzoekwe and Oghosanine (2011); Mustapha et al. (2013) and Mustapha et al. (2015) have reported the presence of nitrogen compounds in petroleum contaminated wastewater. For example, Moreno et al (2002) reported ammonia concentrations ranging from 3 to 20 mg N/L in oil refinery wastewater, while Huddleston et al. (2000) reported ammonia concentrations ranging from 2.14 to 8.6 mg/L in pre-treated petroleum refinery effluent. In addition, nutrients are produced in wastewater due to the degradation of organic matter (Valderrama et al. 2002).

Nitrogen and phosphorus compounds in discharged wastewater may adversely contribute to eutrophication, depletion of oxygen and toxicity to humans, aquatic life and bacteria (Moreno et al. 2002; Abd-El-Haleem et al. 2003; Saeed and Guangzhi 2012) as well as acceleration of the corrosion of metals and construction materials (Taneva 2012). For instance, ammonia is one of the products resulting from purification of oil, 5.0 mg/L of ammonia is the maximum amount allowed in discharged oil refinery effluent according to the Brazilian Environmental Legislation limits (Coneglian et al. 2002). Ammonia creates a large oxygen demand in receiving waters (Abd-El-Haleem et al. 2003), and more than 45 mg/L of nitrate in drinking water may cause methemoglobinemia, while over 8 mg/L NO_2^- inhibits anoxic phosphate uptake (Abd-El-Haleem et al. 2003).

Phosphorus is essential for plant growth and a limiting factor for vegetative productivity (Rani et al. 2011). However, it is also important to remove phosphorus from wastewater since it is a major limiting nutrient for algae growth in fresh water ecosystems (Rani et al. 2011) and thus results in eutrophication of the receiving water bodies (Mulkerrins et al. 2004).

2.2. Constructed wetlands for treatment of petroleum refining wastewater

2.2.1. Conventional treatment technologies

Petroleum related industries have to comply with strict regulations by regulatory authorities. Besides, the concerns over discharge of produced water and reuse of wastewater are the major challenges facing these industries (Allen 2008; Shpiner et al. 2009a). These have led them to explore many treatment technologies as alternative methods for effluent management. Thus, it has expedited the technological advancements in the field of wastewater treatment in the past three decades.

The state-of-the-art technology for the treatment of petroleum contaminated wastewater includes reverse osmosis (Murray-Gulde et al. 2003; Mant et al. 2005), membrane filtration techniques (Allen 2008; Ravanchi et al. 2009) such as microfiltration, ultrafiltration and nanofiltration (Saien 2010), natural sorbent zeolites (Mazeikiene et al. 2005), laccase and peroxidase enzymatic treatment and electrocoagulation (Abdelwahab et al. 2009), air floatation, chemical oxidation (Zhao and Li 2011), microbial degradation (Idise et al. 2010), wastewater stabilization ponds (Shpiner et al. 2009a; Shpiner et al. 2009b) and lagoons (Abdelwahab et al. 2009). Some of these technologies require high energy and a large capital investment (Baskar 2011). The technologies can lead to incomplete decomposition of contaminants (Das and Chandran 2011). Also, some of these methods are basically transfer of contaminants from one medium to another; and this process may require elimination of organic compounds by another treatment method (Diya'uddeen et al. 2011) or otherwise such systems are only suited for primary treatment of produced water (Shpiner et al. 2009a). Furthermore, the by-products of conventional method are often toxic to both humans and the environment (Ojumu et al. 2005; Wuyep et al. 2007; Baskar 2011). Toxic by-products can destroy the biological component of the soil, and can change the chemical and physical soil characteristics (Tam and Wong 2008). Therefore, the by-products require further treatment as well (Prasad 2003).

Other disadvantages of conventional treatment systems include operational difficulties associated with wastewater flow rate and pollutant load (Ayaz 2008), complex procedures, poor performances and high management requirement without oil recovery (Zhao and Li 2011).

2.2.2 Constructed wetland design

CWs are an example of phytoremediation (plant-assisted) systems that have been used for many years to treat wastewater (Yang et al. 2006; Dipu et al. 2010; Tromp et al. 2012). They are low-cost, have a low energy-consumption, low-maintenance, are easily operated and they provide effective treatment (Cheng et al. 2002; Xia et al. 2003; Wallace and Kadlec 2005; Eke and Scholz 2008; Mena et al. 2008; Fountoulakis et al. 2009; Mustapha et al. 2011; Spacil et al. 2011). CWs are chosen to treat many types of wastewater owing to their simplicity, low sludge production, high nutrient absorption capacity, process stability and its potential for creating biodiversity (Rani et al. 2011). In comparison, CWs have the advantages of providing a less intrusive approach than the conventional methods (Lin and Mendelssohn 2009), CWs are environmentally friendly processes where living plants can be considered as a solar driven pump for the extraction of pollutants from wastewater (Wenzel 2009). CWs are economically viable options for wastewater management, especially for developing countries with limited water resources and means (Mustapha et al. 2015). CWs are complex ecosystems (Maine et al. 2007; Fountoulakis et al. 2009) comprising wetland vegetation, hydric soils, microorganisms and prevailing flow patterns that assist in treating wastewater (Haberl et al. 2003; Fountoulakis et al. 2009; Vymazal 2010; Mustapha et al. 2015).

There are two main types of CWs (Fig. 2.1): (i) Surface flow wetlands (those with water flowing above the substrate) that mimic natural wetlands (Fig. 2.2a) both in structures and mechanisms and (ii) subsurface flow wetlands (those with wastewater flowing through the gravel bed or porous media). The latter are further grouped into horizontal subsurface flow (HSSF) (Fig. 2.2c) and vertical subsurface flow (VSSF) wetlands (Fig. 2.2c). Based on the flow direction, VSSF CWs can be further categorized into two types upward flow type and the downward flow type (Stottmeister et al. 2003; Vymazal 2007) and tidal flow (Babatunde et al. 2008). Besides single (surface, subsurface - HSSF and VSSF) CWs, combined (hybrid) CWs have been used to successfully treat various types of contaminants from petroleum-contaminated wastewater (Wallace and Kadlec 2005; Ji et al. 2007; Eke and Scholz 2008; Davis et al. 2009; Aslam et al. 2010).

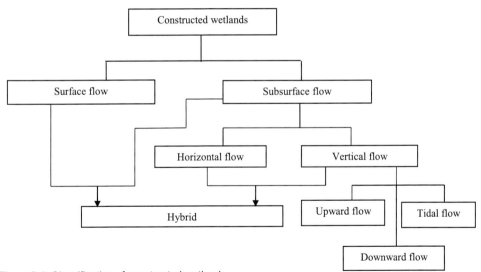

Figure 2.1: Classification of constructed wetlands

2.2.2.1 Constructed wetlands for petroleum refinery wastewater

The petroleum industry has shown an increasing interest in applying CWs to manage wastewater at various sites, including refineries, oil and gas wells and pumping stations (Ji et al. 2007; Allen 2008). Further, the refinery wastewater can be treated with CWs and reused for other purposes rather than the practice of some refineries of pumping the water back into the aquifers at high pressure which is energy intensive and expensive (Shpiner et al. 2009a). In addition, CW systems protect resources for future generations by preserving the ecosystem as well as protecting biodiversity (Schröder et al. 2007).

In spite of the potential to use CWs for the treatment of petroleum contaminated wastewater, literature is scarce (Diya'uddeen et al. 2011) for studies in the tropics. Phytoremediation is favorable in the tropics because of suitable climatic conditions which support plant growth and microbial activity (Merkl et al. 2006). Moreover, countries located in the tropics (Venezuela, Nigeria, and Indonesia) and subtropics (e.g. Saudi Arabia, Kuwait) consists of approximately 80% oil resource (Merkl et al. 2006). In contrast, most of the research on plant-assisted remediation of petroleum-contaminated water is rather conducted in the temperate regions (Merkl et al. 2006).

Thus, there is need for this research to be channeled towards the tropics and subtropics regions. This paper reviews literature specifically on the use of CW systems for treatment of petroleum contaminated wastewater, addressing the efficiency of the treatment systems based on application, as well as organic and inorganic pollutant composition of the influent and effluent, and focusing on the three main components of CWs: macrophytes, microorganisms (found at the petroleum-contaminated sites and in the wastewater) as well as the substrate/filter media used.

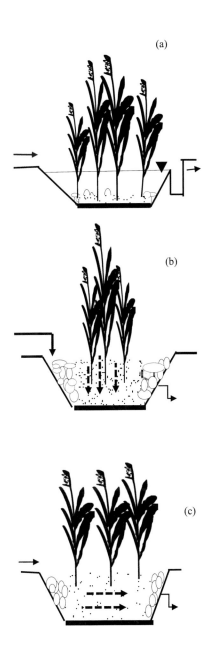

(a)

:

(b)

(c)

Figure 2.2: (a) Surface flow constructed wetland; (b) Vertical subsurface flow constructed wetland and (c) Horizontal subsurface flow constructed wetlands

2.2.2.2 Potential of CWs to treat petroleum wastewater

The application of the different types of CWs for wastewater treatment is dependent on the objectives of the treatment, they can be used as either primary or as secondary or tertiary treatment (Haberl et al. 2003). For instance, subsurface flow (SSF) CWs are preferred as a main treatment system (Haberl et al. 2003) while for tertiary treatment, both surface flow (SF) and SSF CWs can be used. In addition, depending on the desired objectives, HSSF or VSSF CWs are used. The limited oxygen in HSSF CWs provide good conditions for organic matter and denitrification processes (Jakubaszek and Saecka 2015), while COD and ammonia removal will be higher in the VSSF CWs due to the better oxygen transfer capacity through the feeding mode (Lv 2016). Although other mechanisms such as microbial processes, substrate sorption and plant uptake should be considered as well.

2.2.2.2.1 Surface flow CWs

Surface flow (SF) CWs have the ability to filter, absorb and retain particulate matter, nutrients or other pollutants from wastewater (Rani et al. 2011). One difference between SFCWs and SSF CWs is in the release of oxygen. Accordingly, Vymazal (2013) reported that the release of oxygen in SFCWs is insignificant, as most of the treatment processes occur in the water column and within the bottom layer. However, some researchers have used SF CWs to treat petroleum contaminants and have shown this to be a promising technological alternative for wastewater treatment. Horner et al. (2012) used a free-water surface (FWS) flow CW planted with *Typha latifolia* operating at a 4-day HRT under the prevailing conditions at sub-Saharan Africa to treat produced oilfields water containing 0.08 – 0.40 mg/L Fe, 0.50 – 1.26 mg/L Mn, 0.37 – 1.44 mg/L Ni, 2.0 - 5.0 mg/L Zn, 704 – 1370 mg/L total dissolved solids and at varied oil and grease (O&G) concentrations of 10, 25 and 50 mg/L. The FWS CWs has an ability to remove Fe (0-89.2 %), Mn (88.3-98.0%), Ni (23.1-63.2%) and Zn (11.5-84%) while its O&G removal was below the detection limit of 1.4 mg/L (Table 2.2). The effluent concentrations of O&G, Fe and Mn met the criteria for irrigation and livestock watering, while Ni concentrations met the livestock watering criteria. Further, Ji et al. (2007) used SF CWs to treat produced wastewater in the Liaohe oilfield (China) using two reed beds (reed bed # 1 and bed # 2) (Table 2.2). There were variations in the removal efficiency of the two reed beds, these differences may be due to the different hydraulic retention times of 15 and 7.5 d, respectively.

Simi (2000) conducted a study on water quality assessment of a SF CW used for polishing BP Oil's Bulwer Island refinery wastewater in Australia, specifically for reducing the suspended solid loads associated with algal growth in the upstream wastewater treatment system (Table 2.2). The system had low removal rates, which were likely linked to the growth of algae in areas of open water with poor plant establishment. Simi (2000) therefore emphasized the importance of macrophyte selection for system optimization.

2.2.2.2.2 Subsurface flow CWs

Subsurface flow (SSF) CWs are reliable treatment systems with high treatment efficiencies for the removal of pollutants (Hoffman et al. 2011). Davis et al. (2009) used SF and SSF engineered wetlands to treat petroleum hydrocarbons. They reported that SSF engineered wetlands can achieve more biological treatment due to the higher specific surface area present in a gravel bed. In operating a SSF CW, there is no contact between the water column and the atmosphere, this is safer from a public health perspective (Rani et al. 2011).

2.2.2.2.2.1 Vertical subsurface flow CWs

Petroleum wastes are recognized to naturally degrade in natural wetland environments (Wallace and Kadlec 2005; Eke and Scholz 2008; Davis et al. 2009; Wallace et al. 2011). Eke and Scholz (2008) studied twelve vertical-flow microcosm wetlands with different composition to remove benzene (Table 2.2). Their results indicated that the systems had a mean removal efficiency of 85%, which was predominantly due to the volatilization with a one-day retention time. They have, however, concluded that optimizing the wetlands by locating them in areas of high temperature would enhance the biodegradation rates. Eke and Scholz (2008) and Davis et al. (2009) confirm that aeration enhances volatilization and hydrocarbon degradation.

Mustapha et al. (2015) reported removal efficiencies of 43% to 85% for TDS, turbidity, BOD_5, COD, Ammonium-N, Nitrate-N and Phosphate-P respectively, from *C. alternifolius* planted VSSF CWs and 42% to 82% for *C. dactylon* planted VSSF CWs (Table 2.2). *C. alternifolius* and *C. dactylon* planted VSSF CWs were capable of treating secondary refinery effluent to discharge permit limits. According to Vymazal (2010), VSSF CWs provide a more effective removal of organics, suspended solids and ammonia while HSSF CWs provide a higher removal of nitrate through the denitrification processes under anoxic/anaerobic conditions as the concentration of dissolved oxygen in the filtration beds is limited. On average, COD is reduced less effectively than BOD_5 in CWs (Baskar 2011). Moreno et al. (2002) obtained above 90% removal efficiencies for high ammonia inflow concentrations (>

6 mg N/L) from *Phragmites australis* planted vertical up flow CWs with an HRT of 5 h. VSSF CWs have successfully been employed to remove ammonia in refinery wastewater. For instance, Wallace and Davis (2009) reported 98% efficiency of ammonia reduction after treating wastewater that contained more than 230 mg/L ammonia while a similar result has also been reported by Xia et al. (2003). Aslam et al. (2010) assessed the viability of treating refinery wastewater using VSSF CWs filled with coase sand for the removal of heavy metals at a hydraulic loading rate (HLR) of 1.21 m^3/m^2 day. They achieved removal efficiencies of 49% for Fe, 53% for Cu and 59% for Zn.

CWs can be intensified with forced aeration to enhance aerobic biodegradation rates of petroleum hydrocarbons in contaminated wastewater to non-detectable levels (Davis et al. 2009). Wallace (2001) used a 1486 m^2 SSFCW with a Forced Bed Aeration system built into the wetland bed to treat high-strength petroleum contact waste containing 10,000 mg/L carbonaceous biological oxygen demand ($CBOD_5$), 100 mg/L ammonia and 1000 µg/L total BTEX. The SSFCW treatment system achieved average removal rates of 99% for $CBOD_5$, 98% for ammonia and BTEX was removed to non-detectable levels. They reported that BTEX removal was largely due to enhanced volatilization as a result of the aeration system. Similarly, Ferro et al (2002) built a pilot scale SSF wetland system at the BP Amoco refinery site in Wyoming (USA) to treat recovered groundwater contaminated with petroleum hydrocarbons with forced subsurface aeration at influent benzene concentrations that ranged from 0.2 to 0.6 mg/L. Their system consisted of four treatment cells packed with sand and operated in an upward vertical flow mode with a one day mean HRT. The aerated upward vertical SSFCW achieved an effluent with benzene concentrations less than 0.05 mg/L. Wallace and Kadlec (2005) constructed an engineered wetland system with aeration which enhanced the volatilization and aerobic biodegradation. These systems were built at pilot scale and later at full scale operating at 6000 m^3/day and achieved permit compliance within one week after start up. These studies are significant because they demonstrate successful treatment of petroleum compounds with an inbuilt aeration system (Wallace 2001). Bedessem et al. (2007) used a pilot system consisting of four SSF CWs operated in parallel in different flow modes: upward vertical flow or horizontal flow with or without aeration for the treatment of refinery affected groundwater. The treatment system effectively removed total petroleum hydrocarbon-diesel range organics (TPH-DRO) (77% without aeration and > 95% with aeration), total benzene, toluene, ethylbenzene, and o-, m-, and p-xylenes. The systems showed 77% removal for TPH-DRO without aeration and > 95% with aeration, 80% for benzene and 88% for total benzene. However, the treatment system was not effective in reducing the MTBE concentrations. Results from this pilot study were instrumental in the development of an aesthetically pleasing full-scale wetland system for volatile organics removal with cascade aeration pre-treatment.

2.2.2.2.2.2 Horizontal subsurface flow CWs

A couple of studies have shown the success of the removal of organic and inorganic contaminants from petroleum wastewater using HSSF CWs (Table 2.2). However, most of the studies were focused on contaminated groundwater (Ferro et al. 2002; Wallace and Davis 2009; Jechalke et al. 2010; Seeger et al. 2011; Wallace et al. 2011; Chen et al. 2012; Stefanakis et al. 2016). Stefanakis et al. (2016) tested three pilot scale HSSF CWs for the removal of phenolic compounds, MTBE, benzene and ammonia from contaminated groundwater in a pump-and-treat remediation research facility in Germany. The results showed a complete removal of phenol and *m*-cresol and a removal efficiency that varied from 49.6 - 52.8% for MTBE, 72.3 - 82.2% for benzene and about 40% for ammonia. Stefanakis et al. (2016) pointed that microbial processes (nitrification, denitrification) dominated the transformation and removal of ammonium in HSSF CWs, while direct plant uptake is of secondary importance in the long-term.

Davis et al. (2009) presented three field-scale applications of HSSF engineered wetlands with Forced Bed Aeration in North America for pipeline terminal wastewater containing benzene, toluene, xylene and ethylene (BTEX) and ammonia, along with two former refineries using HSSF engineered wetlands with a designed flow rate of 1.5 m^3/d. It treated 6,000 m^3/d of BTEX and 1,060 m^3/d Fe from BTEX-contaminated extracted groundwater, respectively. The systems reduced benzene concentrations from 300 $\mu g/L$ to < 10 $\mu g/L$ (to non-detectable concentrations) at 40% and 80% of gravel bed length, respectively. The high rate of removal was due to enhanced volatilization as a result of the aeration system.

Al-Baldawi et al. (2014) investigated the optimum conditions for total petroleum hydrocarbon (TPH) removal from diesel-contaminated water with *Scirpus grossus* planted in HSSF CWs. Three operational variables were investigated, i.e. diesel concentration (0.1, 0.175 and 0.25% Vdiesel/Vwater), aeration rate (0, 1 and 2 L/min) and retention time (14, 43 and 72 days). They reported that the optimum conditions were found to be a diesel concentration of 0.25 % (Vdiesel/Vwater), a retention time of 63 days and no aeration with an estimated maximum TPH removal from water of 76.3%. This showed that a longer retention time has a positive effect on the reduction of the TPH concentration in water, although the diesel concentration and aeration rate did not have much effect on the TPH removal efficiency. Chen et al. (2012) evaluated the performance of planted, unplanted and plant root mat pilot scale HSSF CWs in the decontamination of groundwater polluted with benzene and MTBE. They reported that the plant root mat showed a similar treatment efficiency as the planted HSSF-CW for benzene removal and a higher treatment efficiency for MTBE removal was achieved in summer time. The main removal pathway in this study was oxidative microbial degradation. Ji et al. (2002) demonstrated the use of SSF CWs for heavy oil produced water (Table 2.2). Thus, effective degradation of organic compounds are majorly by bacterial metabolism of both attached and free living bacteria under anoxic and or anaerobic conditions (Vymazal 2010).

Several authors have demonstrated heavy metal removal by CWs. Though, the application of HSSF CWs for metal removal as the main focus of treatment is rather limited (Kröpfelova et al. 2009). Using aerated and non aerated HSSF CWs planted with *Phragmites australis*, Mustapha et al. (2011) reported that Cd, Cr, Pb and Zn were removed from simulated refinery wastewater passing through the wetland systems with a 2-day HRT at a hydraulic loading rate of 11 L/day (Table 2.2). There were no large variations in the removal efficiency between the aerated and the non-aerated treatment systems (Mustapha et al. 2011). This may likely be due to the low influent diesel concentrations and a two days retention time may be too short to bring about a high difference between the treatments systems. Gillespie et al. (2000) investigated the transfer and transformation of Zn in a refinery effluent. They used two pilot-scale CWs in parallel consisting of an alluvial flood plain sediment planted with *Scirpus californicus* operated at a 24 h nominal HRT. An average of 38% of the total recoverable and 65% of the soluble Zn was removed from the refinery effluent during the experiment.

2.2.2.2.3 Hybrid CWs

Hybrid CWs are non-conventional CWs in which either two or more CWs are combined in series. Thus, hybrid CWs provide a better effluent quality than that of single CW systems. In general, hybrid CWs combine either a VF CW at the first stage with a HF CW at the second stage or vice versa to treat effluents in an efficient manner. For instance, they combine their various advantages to compliment processes that produce an effluent lower in BOD_5 and total nitrogen concentration (Vymazal 2005). The use of hybrid CWs (horizontal + vertical flow or vertical + horizontal flow) is an effective wastewater treatment method with reduced water loss potential (Melián et al. 2010).

Wallace et al. (2011) designed hybrid CWs to reduce metals and organic contaminants as well as buffer the pH of the recovered groundwater (Table 2.2). The Wellsville system was also very effective in iron removal, removing 98% of the iron despite relatively high influent concentrations (mean value of 33.3 mg/L). The high performance of the Wellsville wetland, shows that CWs are a viable and cost-effective treatment alternative to mechanical treatment, even under cold climate conditions.

Kanagy et al. (2008) designed, built and used a modular pilot-scale CW (freshwater wetland and saltwater wetland) to treat four simulated waters (fresh, blackish, saline and hypersaline waters) representing the range of contaminant concentrations typical of actual produced waters. Freshwater wetland cells planted with *Schoenoplectus californicus* and *Typha latifolia* were used to treat the fresh and brackish waters while saline and hypersaline waters were treated by saltwater wetland cells planted with *Spartina alterniflora* and by reverse osmosis (RO). Effective removal of cadmium, copper, lead, and zinc was achieved by the pilot-scale system. The same authors reported that, although the metal concentrations met the targeted levels immediately following the treatment of saline and hypersaline waters by RO, pH levels were typically too low for discharge. Also, during the flow through the freshwater wetland cells, the pH increased to acceptable levels. For all the four types of gas storage produced waters, freshwater wetland cells improved the performance of the system by increasing the dissolved oxygen concentrations. This is because the freshwater wetland was planted with *T. latifolia*, which has the ability to oxygenate its root zone, thus supporting an oxidizing enviroment.

Murray-Gulde et al. (2003) considered conductivity, total dissolved solids (TDS), and toxicity as parameters of concern in their study, *Ceriodaphia dubia* and *Pimephales promelas* were used for the toxicity tests. No significant mortality was observed at 100% exposure to treated produced water when compared to the control organisms. Nonetheless, the system effectively decreased conductivity and TDS by 95% and 94%, respectively.

Plants in CWs can adsorb and accumulate metals. Cheng et al. (2002) used a twin-shaped CW comprising of a vertical flow (inflow) chamber planted with *Cyperus alternifolius*, followed by a reverse-vertical flow (outflow) chamber planted with *Villarsia exaltata* to assess the decontamination of artificial wastewater polluted with Cd, Cu, Pb and Zn for over 150 days and with Al and Mn for 114 days. Heavy metals were undetected in the treated effluent with the exception of Mn. The inflow chamber was, therefore, seen as the predominant decontamination step of more toxic metal species with final concentrations far below the WHO drinking water standards. The lateral roots of *C. alternifolius* accumulated more than 4500 times higher amounts of Cu and Mn from the applied influent, and 100 - 2200 times the amounts of the other metals studied.

Table 2.2. Constructed wetlands for petroleum wastewater treatment

Types of CW	Wastewater type	Range of removal (%)	Location	Reference
SURFACE FLOW				
Surface flow	Leaks underground petroleum storage	Benzene: 48	Diliman, Quezon City, Philippines	Ballesteros Jr. et al. 2016
Surface flow	Produced oilfield water	Fe: NR-89.2, Mn: 88.3-98, Ni:23.1-63.2, Zn:11.5-84, O&G:ND	Clemson, USA	Horner et al. 2012
Surface flow	Heavy oil-produced water	COD:71-80; BOD$_5$:92-93; TKN: 81-88; TP: 81-86	Beijing, China	Ji et al. 2007
Surface flow	Refinery wastewater	BOD$_5$: 10.24; COD:16.44; SS: 14.2; TKN: 14.36; NH$_4$-N: 1.32; TP; 13.44: SRP	Australia	Simi 2000
VSSF				
VSSF CWs	Refinery wastewater	BOD$_5$: 68-70; COD:63-65 NH$_4$+-N:49-68%; NO$_3$-N: 54-58; PO$_4$^{3}+P:42-42	Kaduna, Nigeria	Mustapha et al. 2015
Vertical-flow soil filter systems-Rough filter (RF) Polishing filter (PF) RF + PF (combined)	Refinery wastewater	MTBE: 70, benzene:98 MTBE: 99, benzene:100 MTBE: 100, benzene: 100	Leipzig, Germany	van Afferden et al. 2011
VSSF CWs	Hydrocarbon-contaminated wastewater	Benzene: 85-95	Edinburgh, UK	Eke and Scholz 2008
VSSF CWs	Oil refinery wastewater	Ammonia: 97.7; COD: 78.2; BOD$_5$: 91.4 ; oil: 95.35	Guangzhou, China	Xia et al. 2003
Compost-based and gravel-based vertical flow wetlands HSSF	Refinery wastewater	TSS:51-73 and 39-56 ; COD: 45-78 and 33-61; BOD$_5$: 35-83 and 35-69	Rawalpindi, Pakistan	Aslam et al. 2007

System	Wastewater type	Removal efficiency	Location	Reference
HSSF CWs	Petroleum contaminated wastewater	MTBE: 49.6 - 52.8%; benzene: 72.3 -82.2%; ammonia:40%	Leipzig, Germany	Stefanakis et al. 2016
HSSF CWs	Diesel-contaminated wastewater	TPH: 72.5%	Selangor, Malaysia	Al-Baldawi et al. 2014
HSSF CWs: (1) Unplanted (2) planted (3)plant root mat	Groundwater contaminated with benzene and MTBE	(1) Benzene: 0 -33%; MTBE: 0-33% (2) Benzene: 24 - 100%; MTBE: 16 -93% (3) Benzene:22 -100%; MTBE: 8-93%	Leipzig, Germany	Chen et al. 2012
HSSF CWs	Simulated refinery wastewater – diesel	Cd: 89.9 - 92.5%, Cr: 82.1 - 90.7%; Pb: 84.9 - 90.9% and Zn: 93.8 - 94.2%	Delft, Netherlands	Mustapha et al. 2011
HSSF CWs: (1) planted gravel filter; (2) plant root mat	Groundwater contaminated	(1): Benzene: 81%, MTBE: 17%, NH_4^+-N: 54% (2): Benzene: 99%, MTBE: 82%, NH_4^+-N: 41%	Leipzig, Germany	Seeger et al. 2011
HSSF CWs	Produced water	To non-detect concentration at 40 and 80 of the gravel bed length	USA	Davis et al. 2009
HSSF CWs	Refinery effluent	Total Zn recoverable: 38%; soluble Zn: 65%	Houston, USA	Gillespie et al. 2000
HYBRID 3 detention basins, oil/water separator, a pair of saltwater wetland cells in series, a reverse osmosis unit, and 2 series of four freshwater wetland	Four categories of produced water: (1) fresh (2) brackish (3) saline (4) hypersaline	**Fresh**: Cd: 25%, Cu: ND, Pb: ND, Zn:96.3%, Cl: NR **Brackish:** Cd: 39%, Cu: 89%, Pb: 93, Zn:40%, Cl: 12% **Saline:**	Clemson, USA	Kanagy et al. 2008

cells

Treatment system	Water/Contaminant	Removal efficiency	Location	Reference
		Cd: 99.6%, Cu: 98.8, Pb: 97.7, Zn:99%, Cl: 99% **Hypersaline:** Cd: 99.6%, Cu: >99.9%, Pb: 99.3%, Zn:99.8%, Cl: 99.5%		
Hybrid reverse osmosis-constructed wetlands treatment system	Brackish oil field produced water	Conductivity: 95 ; TDS 94	Clemson, USA	Murray-Gulde et al. 2003
Aerated systems	Hydrocarbon-contaminated	BTEX100%, Aniline 94%, nitrobenzene 93%, Fe 98%	Wellsville, New York, USA	Wallace et al. 2011
Upward VF, HF, with aeration, without aeration-Forced subsurface aeration	Petroleum refinery contaminated groundwater	Fe, benzene, MTBE, TPH(DRO): 77%, BOD_5, TSS, TDS, alkalinity Aerated: Fe, total BTEX: 72.6-85.3%, MTBE, TPH(DRO):94-96%, BOD_5, TSS, TDS, alkalinity	Laramie, Wyoming, USA	Bedessem et al. 2007
Upward vertical subsurface flow wetland system with forced subsurface aeration	Groundwater contaminated with petroleum hydrocarbons	Effluent with benzene concentrations < 0.05 mg/L	Wyoming, USA	Ferro et al. 2002
Subsurface flow constructed wetland (SSFCW) with Forced Bed Aeration system built into the wetland bed	High-strength petroleum contact waste	99% for $CBOD_5$, 98% for ammonia and BTEX was removed at the 40% of the bed length to non-detect level	South Dakota, USA	Wallace 2001

2.3. Removal pathways in constructed wetlands

Constructed wetlands use natural geochemical, physical and biological processes (Fig. 2.3) in a wetland ecosystem to treat contaminants of concern. The bioremediation of contaminants takes place during the passage of raw or pretreated wastewater through the gravel layer and root zone of the constructed wetlands (Babatunde et al. 2008; Kadlec and Wallace 2009). The constituents of concern are removed by various mechanisms such as filtration and sedimentation of suspended particles (Shelef et al. 2013), adsorption to suspended matter, photolysis, volatilization, plant uptake (Zhang et al. 2011) and precipitation by biogeochemical processes (Barber et al. 2001; Stottmeister et al. 2003; Grove and Stein 2005; Farooqi et al. 2008; Shelef et al. 2013). The removal of contaminants occurs by microbial degradation, by physical and chemical processes in aerobic, anoxic as well as anaerobic zones (Vymazal and Brĕzinová 2016). The rhizosphere is the active zone where physicochemical and biological processes occur through interactions between plants, microorganisms and substrates to remove pollutants from wastewater (Khan et al. 2009; Saeed and Guangzhi 2012; Papaevangelou et al. 2017). The major removal mechanisms of organic matter are volatilization, photochemical oxidation, sedimentation, sorption and microbial degradation by fermentation, aerobic and anaerobic respiration (Haberl et al. 2003; Czudar et al. 2011). The mechanisms for nitrogen removal in CWs are nitrification, denitrification, plant uptake, volatilization and adsorption (Rani et al. 2011). The major processes responsible for phosphorus removal in SSFCWs are typically adsorption, precipitation and plant uptake (Rani et al. 2011).

Imfeld et al. (2009) reviewed the various mechanisms contributing to organic matter removal and the main degradation pathways for different groups of contaminants. The removal efficiency of the pollutants is influenced by a number of factors, including the substrate media, redox potential, loading rate, retention time, carbon source availability, electron acceptor concentrations, temperature and the plant species (Zhang et al. 2011).

Reddy and D' Angelo (1997) in their study on the biogeochemical processes of wetlands reported that CWs remediate pollutants because they sustain a number of aerobic and anaerobic biogeochemical processes that control the removal and/or retention of pollutants present in wastewater. These natural processes are also employed to treat petroleum refinery wastewaters (Knight et al. 1999; Xia et al. 2003; Ji et al. 2007; Spacil et al. 2011; Chapter 4). They showed a good potential as a treatment method capable of removing organic and inorganic pollutants from petroleum refinery effluents (Campagna and da Motta Marques

2000; Cheng et al. 2002; Diya'uddeen et al. 2011; Chapter 4; Chapter 6). Yet, the efficiency of pollutant removal mechanisms depends on the hydraulic conductivity of the substrate, types and number of microorganisms, oxygen supply for microorganisms, chemical composition of substrate and hydraulic retention time (Barber et al. 2001; Haberl et al. 2003). For petroleum hydrocarbon removal, design parameters include the biodegradation rate coefficients, flow rate, hydraulic retention time (HRT), influent and required effluent concentrations (Davis et al. 2009).

The main route of heavy metal uptake in aquatic plants is through the roots in the case of emergent and surface floating plants, whereas roots as well as leaves take part in removing heavy metals and nutrients in submerged plants (Dhote and Dixit 2009). Besides, the removal mechanisms in HSSF CWs include binding to sediments and soils through sedimentation, flocculation, adsorption, cation and ion exchange, complexation, oxidation and reduction, precipitation and co-precipitation as insoluble salts and plant uptake and, to a lesser extent, microbial metabolism (Galletti et al. 2010). Numerous factors can affect the remediation processes of contaminated sites, including pH level of water and sediment, mobilization and uptake from the soil, compartmentalization and sequestration within the root, efficiency of xylem loading and transport (transfer factors), distribution between metal sinks in the aerial parts, sequestration and storage in leaf cells, and plant growth and transpiration rates (Hadad et al. 2006; Khan et al. 2009).

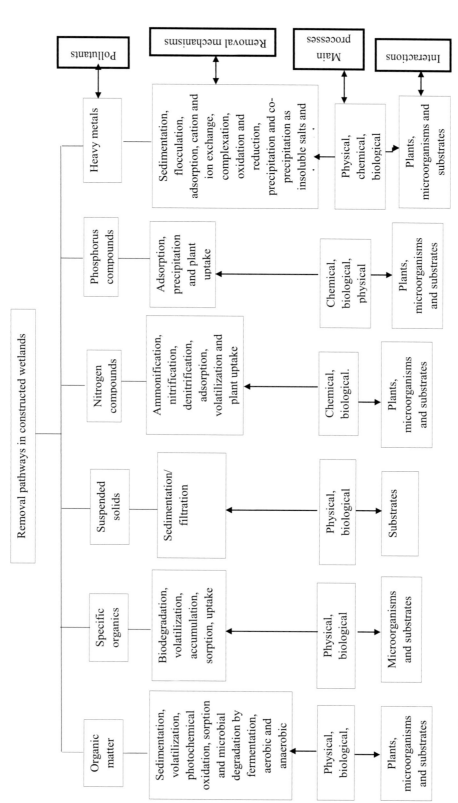

Figure 2.3: Removal mechanisms in constructed wetlands for petroleum wastewater treatment

2.4. Components of constructed wetland treatment

Constructed wetlands are made up of four main components: plants, substrate media, microbial biomass (Dordio and Carvalho 2013) and the aqueous phase (Shelef et al. 2013). The sediment and gravel provide nutrients and support to the root zone and plants (Stottmeister et al. 2003). The root zone is the active reaction zone of CWs, where physicochemical and biological processes are induced by the interaction of the pollutants with the plants, microorganisms and soil particles (Stottmeister et al. 2003). CWs degrade or remove the various pollutants, as a result of the synergetic actions of the system components (Stefanakis et al. 2016). However, the ability of CW to purify wastewater depends on naturally occurring physical, chemical and biological processes that take place within the system.

2.4.1 The macrophyte component

Typically, macrophytes are conspicuous components of a wetland (Vymazal 2011). The macrophytes mainly used in SSF CWs are emergent. These are anchored to the substrate with shoots emerging from water to more than one metre in height, examples are *Phragmites australis, Cyperus alternifolius, Cyperus papyrus* and *Typha latifollia. Phragmites australis* is widely used in CWs all over the world due to their productivity, wide distribution and variable wetland plant species in the world (Lee and Scholz 2007; Vymazal and Brězinová 2016). Surface floating plants are another type of macrophytes, examples include duckweed and water hyacinth. The third group is called submerged macrophytes. Examples of this group include water lilies (*Nymphaea sp.*), *Potamogeton sp., Naja peclinata* and *Ceratophyllum* (Haberl et al. 2003; Allen 2008; Liu et al. 2008).

2.4.1.1 Role of macrophytes in wetland treatment

Wetland systems support a dense growth of vascular plants adapted to saturated conditions (Campagna and da Motta Morques 2001). These plants are known to degrade, extract, contain, or immobilize contaminants in soil and water (Chorom et al. 2012). Vegetation is an essential component of the design of a wetland system (Haberl et al. 2003; Lee and Scholz 2007; Vymazal and Brězinová 2016). The plant roots slow the movement of water, creates microenvironments within the water column and provides attachment sites for the microbial community (Haberl et al. 2003; Lee and Scholz 2007).

Macrophytes are assumed to be the main biological component of wetlands (Hadad et al. 2006; Maine et al. 2007). They are important in their role of pollutant removal, nutrient uptake, accumulation of metals (Cheng et al. 2002; Weis and Weis 2004; LeDuc and Terry 2005), transfer of oxygen to the rhizosphere for growth, microorganisms and decomposition of organic matter (Zhang et al. 2010) and renewing the carbon supplies of metabolizing bacteria (O'Sullivan et al. 2004). Other vital roles played by macrophytes in the efficiency of a CW include flow velocity reduction (which aids in settling of particulates and adsorption of solutes), transportation of gases and solutes between above-ground and below ground biomasses, uptake of inorganic compounds and organic pollutants, and influence the microbial diversity and activity (Taylor et al. 2011). Some plant species can also be used for phytoextraction of heavy metals from contaminated water (Cheng et al. 2002). According to Lee and Scholz (2007), macrophytes have a negative impact on wetland management when they lose their leaves in fall, this will increase the BOD_5 concentration due to the release of carbon, nutrients and other pollutants as well as heavy metals in the litter zone.

Macrophytes have the ability to improve the bioremediation process through diffusion of oxygen from the shoots to the roots as well as to the soil, for soil microbes to utilize it for aerobic respiration (Dowty et al. 2001). This, at times corresponds to as much as 90% of the total oxygen entering a wetland substrate (Allen et al. 2002). Wetland macrophytes have unique characteristics of adaption to anaerobic soil conditions, such as developing internal air spaces (aerenchyma) for supporting supply of oxygen into the root zone (Reddy and D'Angelo 1997). Further, wetland plants have an intrinsic capacity to aerate the rhizosphere i.e. these plants can transport approximately 90% of the oxygen available in the rhizosphere (Lee and Scholz 2007), thus potentially increasing both aerobic decomposition of organic matter and the growth of nitrifying bacteria (Lee and Scholz 2007; Lin et al. 2009).

Wetland plants can incorporate pollutants directly into their tissues, act as catalysts for purification reactions by escalating the environmental diversity in the rhizosphere, thus promoting different types of chemical and biochemical reactions that can improve treatment processes (Maine et al. 2007). Additionally, plants can uptake pollutants (including polyaromatic hydrocarbons) into their rhizosphere to varying extents through the transpiration stream (Weyens et al. 2009). Additionally, all plants have the ability to accumulate essential metals from the soil solution for growth and development. This potential allows plants to also take up other non-essential metals like Al, As, Au, Cd, Hg, Pb, Pt, Sb, Te, T, and U, which have no biological function (Jadia and Fulekar 2009).

For wetlands constructed to treat petroleum contaminated wastewater, a number of macrophytes have proven effective in the degradation of contaminants of concern. This is because of their high biomass production, assimilation and long term storage of organic and inorganic pollutants (Campagna and da Motta Marques 2001) as well as their natural ability to treat wastewater contaminated with oil and grease (Xia et al. 2003; Ji et al. 2007; Davis et al. 2009). Table 2.3 presents different types of wetland vegetation used for the treatment organic and inorganic contaminants found in petroleum refining wastewater. The removal of the contaminants varied for different contaminants as well as for the plants (Table 2.3). For natural wastewater treatment systems, plant productivity and pollutant removal efficiency are important in selecting a suitable plant for a given application (Haberl et al. 2003). Accordingly, Madera-Parra et al. (2015) stated that appropriate plant selection is crucial to improve the heavy metal removal efficiency of CWs. Aside for heavy metal removal, Brisson and Chazarenc (2009) are of the opininon that plants species selection is fundamental to the overall pollutant removal efficiency of a CW.

Many studies have reported significantly higher removal efficiencies of pollutants and enhanced transformation of contaminants in planted CWs, compared to unplanted CWs (Tanner 2001; Merkl et al. 2005; Davis et al. 2008; Seeger et al. 2011; Taylor et al. 2011; Vymazal 2011; Noori et al. 2015; Papaevangelou et al. 2017; Cheng et al. 2017; McIntosh et al. 2017). This is attributable to the activity of the microbes in the rhizosphere (Haberl et al. 2003; Seeger et al. 2011; Cheng et al. 2017). Stefanakis et al. (2016) reported the performance for three HSSF CWs: A and C (planted) and B (unplanted) operating under the same conditions. The planted beds achieved better removal efficiencies than the unplanted bed Bwith removal efficiencies of 52.8% and 82.2% for Bed A, 49.6% and 72.3% for bed C and 41.2% and 66.1% for bed B, for MTBE and benzene, respectively. Similarly, in a study by Mustapha et al. (2015), VSSF CWs planted with *Cyperus alternifolius* and *Cynodon dactylon* were significantly more effective than the unplanted VSSF CWs. The planted VSSF CWs were able to reduce the concentrations of contaminants in the Kaduna refinery effluent to the compliance limits set by the World Health Organization (WHO) and the Federal Environmental Protection Agency (FEPA) of Nigeria. Also, from the results of an investigation conducted by Taylor et al. (2010) with monocultures of 19 plant species, the average COD removal in unplanted wetlands was 70% while the same parameter of individual species was range from 70 to 97 %. Also, statistically significant differences in transformation of organic compounds were found in the rhizosphere of plants from the sedge and rush families (*Cyperaceae and Juncaceae*) compared to the grass (*Poaceae*) family (Taylor et al. 2010).

Merkl et al. (2005) reported that macrophytes can improve microbial degradation by supplying oxygen to the root area along loosened soil aggregates and that remediation of petroleum hydrocarbons is based on the stimulation of microbial degradation in the rhizosphere. Chapelle (1999) also stressed the significance of oxygen as well as nitrogen and phosphorus for the biodegradation of petroleum hydrocarbons. For example, Eke and Scholz (2008) used *Phragmites australis* in their experiment for benzene removal. They reported that *Phragmites australis* does not play an important role in removing benzene, in spite of being given supplementary oxygen through its rhizomes, except nutrients (including fertilizer) are provided. Dowty et al. (2001) further noted that if there is insufficient oxygen, microbial degradation of oil may produce compounds that are toxic to plants (such as hydrogen sulphide) or decrease limiting nutrients (nitrogen) to the extent of inhibiting plant growth.

Table 2.3: Examples of macrophytes used in petroleum contaminated wastewater treatment

Macrophytes	Removal efficiency	Reference
Cyperus alternifolius, Cynodon dactylon (L.) Pers.	TDS: 50-54%, BOD: 68-70%, COD: 63-65%, NH_4^+-N: 49-68%, NO_3-N: 54-58% and PO_4^{3-}P: 42-43%	Mustapha et al., 2015
Scirpus grossus	72.5%	Al-Baldawi et al., 2014
Typha latifolia	Fe: 49 %; Cu: 53%; Zn: 59%	Aslam et al., 2010
Bur-reed, bulrush, *Typha latifolia,* dogwood	To non-detect concentration at 40 % and 80 % of the gravel bed length	Davis et al., 2009
Juncus roemerianus	PAH: 84-100 %; n-alkanes : 85-99.8%	Lin and Mendelssohn, 2009
Schoenoplectus californicus, Typha latifolia	Cd:25 - 99.6%, Cu: 89% -ND, Pb:93% - ND, Zn: 40 – 99.8%, Cl: NR – 99.5%	Kanagy et al., 2008
Phragmites karka	TSS:51-73 % and 39-56%; COD: 45-78 % and 33-61 %; BOD: 35-83 % and 35-69 %	Aslam et al., 2007
Phragmites spp.	COD : 71-80 %; BOD : 92-93 %; TKN : 81-88 %; mineral oil: 81-86 %	Ji et al., 2007
Typha latifolia, Scirpus californicus	95 %; 94 %;	Murray-Gulde et al., 2003
Vetiveria zizanioides, Phragmites australis, Typha latifolia, Lepironia artcutala	Ammonia N: 97.7 %; COD:78.2 %; BOD: 91.4 %; oil: 95.35 %	Xia et al., 2003
Phragmites australis	90 %	Moreno et al., 2002
Scirpus californicus	66 % of soluble Zn	Gillespie Jr et al., 2000

Phragmites australis	COD: 10.24 %; SS: 16.44 %; TKN:14.2 %; NH₄⁺-N: 14.36 %; TP: 1.32 %; SRP: 13.44%	Simi, 2000

2.4.2 Microorganisms

2.4.2.1 Microbial ecology of petroleum degrading constructed wetlands

Constructed wetland systems support an ideal environment for the growth of microorganisms (Saeed and Guangzhi 2012) which gives the microorganisms' tremendous potential to uptake and degrade pollutants (Wuyep et al. 2007). Microorganisms play the most crucial role in the transformation and mineralization of nutrients and organic pollutants (Dordio and Carvalho 2013). Indigenous microorganisms utilize petroleum contaminants of crude oil as carbon and energy source, thereby breaking down the hydrocarbons into simple non-toxic compounds such as CO_2 and H_2O (De-qing et al. 2007). Petrohiles are unique microorganisms that utilize these hydrocarbons as energy and carbon source (Bako et al. 2008).

Bioremediation is to a great extent enhanced by higher temperatures, humidity and soil radiation (Merkl et al. 2005). A diverse microbial community of bacteria, fungi, algae, and protozoa present in the aerobic and anaerobic zones of a wetland (Scholz 2003; McIntosh et al. 2017), is able to degrade volatile organics such as benzene, toluene, ethylbenzene and p-xylene (Eke and Scholz 2008; Sepahi et al. 2008; Davis et al. 2009). Dordio and Carvalho (2013) stated that biodegradation of petroleum organic compounds is increased under aerobic condition, while PCBs degrade faster under moderately reduced conditions.

It is assumed that the actual degradation process is performed by microorganisms in the rhizosphere (Merkl et al. 2006). Therefore, their activities are very important to the accomplishment of any treatment process in wetlands (Scholz 2003). Hydrocarbon utilizing bacteria are present in almost all soil types and their population will grow if the right feedstock is available (Ayotamuno et al. 2006; McIntosh et al. 2017). In the rhizosphere, microbial diversity, density, and activity are more abundant and this can promote increased phytoremediation activity (Hietala and Roane 2009). The enhanced remediation of contaminants in the rhizosphere is due to high microbial densities and metabolic activities in the rhizosphere, which can be attributable to microbial growth on root exudates and cell debris originating from the plant roots (Weyens et al. 2009).

To improve the biodegradation rate of hydrocarbons, some activities can be considered: (i) addition of nutrients, (ii) watering and (iii) bioagumentation, i.e. addition of suitable microbiota (De-qing et al. 2007; McIntosh et al. 2017). Some scientists have reported that the action of adapted microorganisms depends on the microbial composition, contaminant type, geology of the polluted site and chemical conditions of the contaminated site (Sepahi et al. 2008). Ayotamuno et al. (2006) investigated the remediation of a crude oil spill by combining biostimulation with agricultural fertilizers. They concluded that enhanced degradation of petroleum hydrocarbons can be achieved by the addition of nutrients and oxygen. This was also confirmed by De-qing et al. (2007) and Sepahi et al. (2008). In addition, Fernandez-Luqueno et al. (2011) concluded in their review that indigenous microorganisms have the potential of remediating PAHs from a contaminated soil, although if the soil lacks nutrients it could hinder microbial activity and consequently the mineralization of these contaminants.

2.4.2.2 Potential of hydrocarbon degrading microorganisms

The complex mixture of hydrocarbons in crude oil hinders their complete degradation by a single strain of microorganisms. Therefore, degradation is mostly achieved by microbial consortia and their broad enzymatic capacity (Merkl et al. 2006; Scullion 2006). Examples of such diverse community members capable of utilizing crude oil as source of carbon and energy include: *Pseudomonas putida, Flavobacterium spp.* (Huang et al. 2005); *Pseudomonas fluorescence* (Ojumu et al. 2005); *Corynebacterium, Micrococcus, Acinetobacter, Aerococcus* (Ayotamuno et al. 2006); *Staphylococcus, Serratia, Chromobacterium, Alkaligenes* (Abu and Dike 2008); *Pseudomonas aeruginosa, Penicillium janthinellum* (Bako et al. 2008); *Mycobacterium parafortuitum, Sphingobium yanoikuyae* (Tam and Wong 2008); B*acillus spp.* (Sepahi et al. 2008); and *Bacillus cereus* (Idise et al. 2010).

Abu and Dike (2008) monitored natural attenuation processes in a model microcosm wetland representing a typical Niger Delta (Nigeria) environment by comparing natural and enhanced processes. A total of 28 bacteria were isolated and identified to the genus level (Table 2.4). These bacteria have been reported by different researchers as crude oil degraders (Ojumu et al. 2005; Bako et al. 2008, Idise et al. 2010). Their results showed a higher prevalence of gram negative rod forms. They noted that naturally occurring microorganisms in crude oil-impacted sediments can utilize hydrocarbons. However, oxygen was a limiting factor for biodegradation of oil in polluted wetlands.

BTEX compounds are highly soluble and are extremely mobile in groundwater (Jechalke et al. 2010). Benzene is the most toxic and water soluble BTEX compound while it can be degraded by many microorganisms under oxic and hypoxic conditions (Jechalke et al. 2010). Table 2.4 summarizes the potential of hydrocarbon degrading bacteria from the literature.

Ojumu et al. (2005) studied *Pseudomonas aeruginosa* and *Pseudomonas fluorescence* for their bioremediation potential of phenol biodegradation in refinery effluent. Phenol was degraded completely by *P. aeruginosa* and *P. fluorescence* within 60 and 84 h, respectively. Bako et al. (2008) evaluated the potential of *P. aeruginosa* and *P. janthinellium* and their mutants had effective degradation of crude oil in the water samples taken from River Kaduna (Nigeria) after two weeks of incubation at 30 ^{0}C. The study of Agarry et al. (2008) revealed a high potential of *P. aeruginosa* NCIB 950 and *P. fluorescence* NCIB 3756, with *P. aeruginosa* being more effective in degrading phenol in refinery effluent. Idise et al. (2010) emphasized the need to stimulate the strains with organic fertilizer to achieve a better outcome. *Bacillus* spp. (*Bacillus S6* and *Bacillus S35*) isolated from crude oil was able to utilize crude oil as carbon and energy source with increased optical densities and total viable count concomitant with a pH decrease on the fifth day of the experiment (Sepahi et al. 2008). *Pseudomonas* species are also accountable for about 87% of gasoline degradation in contaminated aquifers (Eke and Scholz 2008).

Weyens et al. (2009) stressed the significance of plant-microbe partnerships for a successful removal of organic contaminants. They also demonstrated how plants depend on their associated microorganisms to efficiently remove organic compounds. These associated microorganisms enhance the capability for a stepwise transformation of organic (petroleum) contaminants by consortia and provide an environment that is favourable for genetic exchange and gene rearrangements (Weyens et al. 2009).

Table 2.4: Effectiveness of strains of microorganisms involved in bioremediation of crude oil polluted medium

Type of Microorganism	Techniques	Comment	Reference
Pseudomonas aeruginosa and *Pseudomonas fluorescence*	Batch fermentation and continuous culture	*Pseudomonas aeruginosa* was able to completely remove phenol from the refinery effluent within 60 h of cultivation and while *Pseudomonas fluorescence* could only remove 73.1% of phenol within the same period.	Ojumu et al., 2004
Pseudomonas aeruginosa and *Penicillium janthinellium*	Incubation for 2 weeks	All consortia were observed to have significant decreases in contents of phenol, oil and grease, phosphates, ammonia, nitrates, and sulphates after two weeks of incubation at 30^{0}C.	Bako, et al., 2008

Pseudomonas, Micrococcus, Flavobacterium, Staphylococcus, Serratia, Corynebacterium, Bacillus, Chromobacterium and *Alkaligenes*	Natural attenuation	Microorganisms occurring naturally in crude oil-impacted sediment utilize hydrocarbons and therefore, remediation of oil polluted environment can be achieved.	Abu and Dike, 2008
Pseudomonas aeruginosa NCIB 950 and *Pseudomonas fluorescence* NCIB 3756	Batch fermentation process	Phenol was successfully degraded by both species; with *P. aeruginosa* more effective. They can both be used for bioremediation of petroleum refinery wastewater.	Agarry et al., 2008
Bacillus cereus and *Pseudomonas aeruginosa*	Organic fertilizer and modified strains	The strains achieved better results (98.25% for oil and grease and 87.34% for total petroleum hydrocarbons) when modified, and treated with organic fertilizer (NPK 15-15-15).	Idise et al., 2010
Bacillus spp.	Growth of isolated *Bacillus* on crude oil	The results of the test revealed that *Bacillus spp.* can utilize crude oil as a carbon and energy source.	Sephahi et al., 2008
Bacillus sp. and *Pseudomonas* sp.	Bioaugmentation with bacteria and biostimulation with poultry manure	Bioaugmentation was more effective than biostimulation.	Ijah and Antai, 2003
Bacillus, Corynebacterium, Pseudomonas, Flavobacterium, Micrococcus, Acinetobacter and *Aerococcus*	Fertilizer application and oxygen exposure	With nutrient supplementation, bioremediation can achieve high rates of degradation of petroleum hydrocarbons in agricultural soils.	Ayotamuno et al., 2006

2.4.3 Role of substrate media of constructed wetlands

The substrate medium is an important component of a CW (Dordio and Carvalho 2013; Papaevangelou et al. 2017). It provides surface area for plant and microbial film growth (Papaevangelou et al. 2017). Furthermore, plant roots and microorganisms can have their supplies of water, air and nutrients as well as some moderation of environmental conditions that can influence their development, such as temperature or pH in the substrate media (Dordio and Carvalho 2013). The substrate medium is the primary sink for heavy metals present in the aquatic environment (Tirkey et al. 2012; Yadav et al. 2012; Papaevangelou et al. 2017) and oil contaminants if they are not degraded (Abu and Dike 2008). Furthermore, both the plant root zone and the substrate absorb ionic heavy metals (Chen et al. 2009) and substrate media can have a higher absorptive capacity than the plant roots (Galletti et al.

2010; Rani et al. 2011). Fine textured sediments can accumulate metals if they contain high amounts of organic matter, in contrast, coarse textured materials have low affinity for metals (Galletti et al. 2010). Papaevangelou et al. (2017) reported that Cr removal was accomplished mainly through the substrate and attached organic matter, rather than plant itself through the phytoremediation processes.

Gravel and sand are the most common types of substrate media used in CWs. These substrate media have a limited adsorptive capacity and the capability of a wetland to remove inorganic pollutants can greatly reduce overtime (Hua et al. 2015). Therefore, the capacity of substrate media can be greatly improved by using active filter materials, e.g. with reactive Fe/Al hydrous oxide adsorption surfaces (Hua et al. 2015).

2.5. Capital, operation and maintenance costs

The investment costs to consider for the construction of wetlands are basically land acquisition, site survey, system design, site preparation, plastic liners for prevention of ground water contamination, filtration, rooting media, vegetation, hydraulic control structures and miscellaneous costs which may include fencing and access roads (Kivaisi 2001; Rousseau et al. 2008; Vymazal 2010). In addition, Haberl et al. (2003) included specific demands and circumstances of the site such as topography, distance to the receiving water, existing devices and availability of necessary area into the costs. The costs of construction greatly differ from one site-specific factor to the other, for example, flow control structures may vary from US $2000 to US$ 80 000/10,000 m^2 for SF wetlands and up to US$ 150 000/10,000 m^2 for subsurface flow (SSF) wetlands (Allen, 2008) and approximately €2200/PE in Upper Austria (Haberl et al. 2003). Excavation costs vary between 7 and 27.4% of the total capital cost, while other cost goes to gravel (27 - 53%), lining (13-33%), plants (2-12%), plumbing (6-12%), control structures (3.1-5.7%) and miscellaneous (1.8-12%) (Vymazal 2010). The total investment costs vary even more globally, and the cost could be as low as US $29/m^2$ in India or US$33/$m^2$ in Costa Rica, or as high as €257/m^2 (Vymazal 2010) and €392/PE for SF and €1,258/PE for SSF wetlands in Belgium (Rousseau et al. 2008).

The capital cost for CWs may be influenced by the choice of substrate, plant species, basin compartmentation, lining, flow structure, and other CWs components (Calheiros et al. 2009). In general, the capital costs for FWS CWs are generally lower than for SSF CWs, this is primarily due to lower quantity of media required for rooting soil on the bottom of the beds

(Vymazal 2010). The average capital costs for SF wetland systems are US$200,000 per hectare, while the free surface flow (FWS) systems cost approximately US$50,000 per 10,000 m^2. The main cost difference between the two systems is in the expenses of acquiring the gravel media and transporting it to the site. The construction cost per hectare is higher for SF wetlands. The unit cost is US$163/m^2 (US$0.21/L) of wastewater treated for the SF type, and US$206/m^2 ($0.21/L) for the FWS type (EPA (1993).

The cost of construction of CWs and that of conventional wastewater treatment plants are similar (Haberl et al. 2003). Thus, well-designed wetland systems have low operating costs, making them economically competitive with conventional treatment systems (Allen 2008). Operation and maintenance (O&M) costs of CW include pumping energy if necessary, compliance monitoring, maintenance of access roads and berms, pre-treatment, weed control, vegetation harvesting, effluent sampling and control, cleaning of distribution systems and pumps, equipment replacement and repairs (Haberl et al. 2003; Rousseau et al. 2008; Vymazal 2010). Rousseau et al. (2008) reported an estimated range of US$2500 and US$5000/ha/yr for O&M costs for SSF CW and about US$1000/ha/yr for median cost for SF CW. The basic costs are much lower, by a factor of 2-19, than those for competing technologies using concrete and steel constructions (Vymazal 2010).

2.6. Conclusions

- Surface and subsurface flow constructed wetlands were successfully utilized for the remediation of petroleum-contaminated wastewater. They simultaneously treated multiples contaminants (BOD$_5$, COD, hydrocarbons, oil and grease, heavy metals, nutrients and suspended solids) in the petroleum wastewater with a high level of efficiency (> 70 %) with oil and grease and benzene treated below detection limits. They are, therefore, a sustainable substitute to mechanical treatment systems to remediate petroleum and oil refinery wastewater.
- The hybrid constructed wetlands showed a removal efficiency > 90% for CBOD$_5$, TPH, BOD$_5$, TSS, BTEX and ammonia and benzene to non-detect level. Also, the aerated constructed wetland systems showed a greater removal efficiency of the contaminants of interest compared with the removal performance by the non-aerated system.
- The ability of constructed wetlands to purify wastewater depends on naturally occurring physical, chemical and biological processes that take place within the system that

remove the various pollutants induced by the interaction of the plants, microorganisms and soil particles.

- Each of the components of constructed wetlands (plants, substrate media, microbial biomass and the aqueous phase) play a crucial role in the transformation and mineralization of nutrients and organic pollutants. The microorganisms play the most vital role. The substrate medium is the primary sink for pollutants present in the aquatic environment, it also have a higher absorptive capacity compared with the plant roots.

- Surface flow constructed wetlands have the potential of breeding pests, odour problems and generally, all the constructed wetlands types require greater land area than the conventional mechanical treatment systems.

- Overall, the remediation processes are enhanced by higher temperatures, humidity and soil radiation which is favoured by the tropical climatic conditions.

2.7. Acknowledgement

The authors acknowledge the Government of the Netherlands (Netherlands Fellowship Programme) for their financial assistance (NFP-PhD CF7447/2011).

2.8. References

Abd-El-Haleem D, Beshay U, Abdelhamid AO, Moawad H, Zaki S (2003) Effects of mixed nitrogen sources on biodegradation of phenol by immobilized *Acinetobacter sp.* strain W-17. Afric J Biotechnol 2(1): 8-12. Retrieved from http://www.academicjournals.org/AJB

Abdelwahab O, Amin NK, El-Ashtoukhy E-SZ (2009) Electrochemical removal of phenol from oil refinery wastewater. J Hazard Mater 163: 711-716. doi:10.1016/j.jhazmat.2008.07.016

Achile GN, Yillian L (2010). Mineralization of organic compounds in wastewater contaminated with petroleum hydrocarbon using Fenton's reagent: a kinetic study. American J Sci 6(4):58-66. Retrieved from http://www.americanscience.org

Adewuyi GO, Olowu, RA (2012) Assessment of oil and grease, total petroleum hydrocarbons and some heavy metals in surface and groundwater within the vicinity of NNPC oil depot in Aata, Ibadan metropolis, Nigeria. IJRRAS 13(1):166-174. Retrieved May 14, 2016, from www.arpapress.com/Volumes/Vol13Issue1/IJRRAS_13_1_18.pdf

Agarry SE, Durojaiye AO, Yusuf RO, Solomon BO, Majeed O (2008) Biodegradation of phenol in refinery wastewater by pure cultures of *Pseudomonas aeruginosa* NCIB 950 and *Pseudomonas fluorescence* NCIB 3756. Int J Environ Pollut 32(1): 3-11. doi:10.1504/IJEP.2008.016894

Al-Baldawi I A, Abdullah SR, Abu Hasan H, Suja F, Anuar N, Mushrifah I (2014) Optimized conditions for phytoremediation of diesel by *Scirpus grossus* in horizontal subsurface flow constructed wetlands (HSFCWs) using response surface methodology. J Environ Manage 140: 152-159. doi:10.1016/j.jenvman.2014.03.007

Ali N, Oniye S, Balarabe M, Auta J (2005) Concentration of Fe, Cu, Cr, Zn and Pb in Makera-Drain, Kaduna, Nigeria. ChemClass J 2:69-73.

Allen WC, Hook PB, Biederman JA, Stein OR (2002) Temperature and wetland plant species effects on wastewater treatment on wastewater treatment and root zone oxidation. J Environ Qual 31(3):1010-1016.

Allen WE (2008). Process water treatment in Canada's oil sands industry: II. A review of emerging technologies. J Environ Eng Sci 7:499-524. doi:10.1139/S08-020

Alsalhy QF, Almukhtar RS, Alani HA (2015) Oily refinery wastewater treatment by using membrane bioreactor (MBR). Arab J Sci Eng 1-11. doi:10.1007/s13369-015-1881-9

Asatekin A, Mayes A (2009) Oil Industry wastewater treatment with fouling resistant membranes containing amphiphilic comb Copolymers. Environ Sci Technol 43(12): 4487-4492. doi:10.1021/es803677k

Aslam MM, Malik M, Braig MA (2010) Removal of metals from refinery wastewater through vertical flow constructed wetlands. Int J Agric Biol 12:796-798. Retrieved from http://www.fspublishers.org

Aslam MM, Malik M, Baig M, Qazi I, Iqbal J (2007) Treatment performances of compost-based and gravel-based vertical flow wetlands operated identically for refinery wastewater treatment in Pakistan. EcolEng 30: 34–42. doi:0.1016/j.ecoleng.2007.01.002

Ayaz S (2008) Post-treatment and reuse of tertiary treated wastewater by constructed wetlands. Desalination 226(1-3): 249-255. doi:doi:10.1016/j.desal.2007.02.110

Ayotamuno MJ, Kogbara RB, Ogaji SO, Robert SD (2006) Bioremediation of a crude-oil polluted agricultural soil at Port Harcourt, Nigeria. Appl Energy 83: 1249-1257.

Babatunde AO, Zhao Y, O'Neill M, O'Sullivan B (2008) Constructed wetlands for environmental pollution control: A review of developments, research and practice in Ireland. Environ Inter 116-126. doi:10.1016/j.envint.2007.06.013

Bako SP, Chukwunonso D, Adamu AK (2008) Bioremediation of refinery effluents by strains of *Pseudomonas aerugenosa* and *Penicillium Janthinellum*. Appl Ecol Environ Res 6(3): 49-60. Retrieved from http://www.ecology.uni-corvinus.hu.

Ballesteros Jr F, Vuong TH, Secondes MF, Tuan PD (2016) Removal efficiencies of constructed wetland and efficacy of plant on treating benzene. Sustainable Environ Res 26: 93-96. doi:10.1016/j.serj.2015.10.002

Barber LB, Leenheer JA, Noyes TI, Stiles EA (2001). Nature and transformation of dissolved organic matter in treatment wetlands. Environ Sci Technol35(24): 4805-4816. doi:10.1021/es010518i

Baskar G, 2011. Studies on application of subsurface flow constructed wetland for wastewater treatment. SRM University, Kattankulathur - 603 203, Department of Civil Engineering . Chennai, Tamil Nadu, India: SRM University, Kattankulathur - 603 203.

Bedessem ME, Ferro AM, Hiegel T (2007) Pilot-Scale Constructed Wetlands for Petroleum-Contaminated Groundwater. Water Environ Res 79: 581-586. doi:10.2175/106143006X111943

Brisson J, Chazarenc F (2009) Maximizing pollutant removal in constructed wetlands: Should we pay more attention to macrophyte species selection? Sci Total Environ 407(13): 3923-3930. doi:10.1016/j.scitotenv.2008.05.047

Brito EM, Duran R, Guyoneaud R, Goñi-Urriza M, Oteyza D, García de Oteyza, T, Crapez MAC, Aleluia I, Wasserman JC (2009) A case study of in situ oil contamination in a mangrove swamp (Rio De Janeiro, Brazil). Marine Pollut Bulletin 58: 418–423. doi:10.1016/j.marpolbul.2008.12.008

Calheiros CS, Rangel AO, Castro PM (2008) The effects of tannery wastewater on the development of different plant species and chromium accumulation in *Phragmites australis*. Achives of Environ ContaminatToxicol 55(3): 404-414. doi:10.1007/s00244-007-9087-0

Calheiros CS, Rangel A, Castro PM (2009) Treatment of industrial wastewater with two-stage constructed wetlands planted with *Typha latifolia* and *Phragmites australis*. Bioresour Technol 100(12): 3205-3215. doi:10.1016/j.biortech.2009.02.017

Campagna A, da Motta Marques D (2000) The effect of habitat heterogeneity in experimental wetlands receiving petroleum refinery effluent on the size and clonal modules location of aquatic macrophytes *Scirpus californicus*, *Typha subulata* and *Zizaniopsis bonariensisw*. *7th International Conference on Wetlands Systems for Water Pollution Control*, pp. 485-493. Florida.

Campagna A, da Motta Morques D (2001) The effect of refinery effluent on the aquatic macrophytes *Scirpus californicus*, *Typha subulata* and *Zizaniopsis bonariensisw*. Water Sci Technol 44(11-12): 493-498.

Chapelle F (1999) Bioremediation of petroleum hydrocarbon-contaminated ground water: The perspectives of history and hydrology. Ground Water 37(1),: 122-132. doi:10.1111/j.1745-6584.1999.tb00965.x

Chen M, Tang Y, Li X, Yu Z (2009) Study on the heavy metals removal efficiencies of constructed wetlands with different substrates. J Water Resour Protect 1: 1-7. Retrieved from http://www.SciRP.org/journal/jwarp/

Chen Z, Kuschk P, Reiche N, Borsdorf H, Kästner M, Köser H (2012) Comparative evaluation of pilot scale horizontal subsurface-flow constructed wetlands and plant root mats for treating groundwater contaminated with benzene and MTBE. J Hazard Mater 209-210: 510-515. doi:10.1016/j.jhazmat.2012.01.067

Cheng L, Wang Y, Cai Z, Liu J, Yu B, Zhou Q (2017) Phytoremediation of petroleum hydrocarbon-contaminated saline-alkali soil by wild ornamental Iridaceae species. Int

J Phytorem 19(3):300-308. Retrieved from http://dx.doi.org/10.1080/15226514.2016.1225282

Cheng S, Grosse W, Karrenbrock F, Thoennessen M (2002) Efficiency of constructed wetlands in decontamination of water polluted by heavy metals. Ecol Eng 18(3): 317-325. doi:http://dx.doi.org/10.1016/S0925-8574(01)00091-X

Chorom M, Parnian A, Jaafarzadeh N (2012) Nickel removal by the aquatic plant (*Ceratophyllum Demersum* L.). IntJ Environ Sci Develop 3(4):372 - 375.

Coneglian CM, Conceição DM, de Angelis D d, Marconato JC, Bidóia E (2002) Concentration reduction of ammonia in industrial oil residue of an oil refinery. *Salusvita, Bauru* 21(1):43-50.

Czudar A, Gyulai I, Kereszturi P, Csatari I, Serra-Paka S, Lakatos G (2011) Removal of organic materials and plant nutrients in a constructed wetland for petrochemical wastewater treatment. *Studia Universitatis "Vasile Goldiş", Seria Ştiinţele Vieţii, 21*(1), 109-114. Retrieved from www.studiauniversitatis.ro

Das N, Chandran P (2011) Microbial degradation of petroleum hydrocarbon contaminants: an overview. Biotechnol Res Int 2011: 1-13. doi:10.4061/2011/941810

Davis BM, Wallace S, Willison R (2009). Pilot-scale engineered wetlands for produced water Treatment. Society of Petroluem Engineers 4(3): 75-79. Retrieved December 22, 2016, from http://dx.doi.org/10.2118/120257-PA

De Biase C, Reger D, Schmidt A, Jechalke S, Reiche N, Martínez-Lavanchy PM, Rosell M, Van Afferden M, Maier U, Oswald, Sascha E, Thullner M (2011) Treatment of volatile organic contaminants in a vertical flow filter: Relevance of different removal processes. Ecol Eng 37: 1292-1303. doi:10.1016/j.ecoleng.2011.03.023

Dhote S, Dixit S (2009) Water quality improvement through macrophytes - a review. Environ Monit Assess 152(1):149-153. doi:10.1007/s10661-008-0303-9

Dipu S, Anju A, Kumar V, Thanga SG (2010) Phytoremediation of Dairy Effluent by Constructed Wetland Technology Using Wetland Macrophytes. Global J Environ Res 4(2): 90-100.

Diya'uddeen BH, Abdul Aziz AR, Wan Daud WM (2012) Oxidative mineralisation of petroleum refinery effluent using Fenton-like process. Chem Eng Res Design 90: 298-307. doi:10.1016/j.cherd.2011.06.010

Diya'uddeen BH, Wan Daud MA, Abdul Aziz AR (2011) Treatment technologies for petroleum refinery effluents: a review. Process Safety Environ Protect 89: 95-105. doi:10.1016/j.psep.2010.11.003

Dordio A, Carvalho A (2013) Organic xenobiotics removal in constructed wetlands, with emphasis on the importance of the support matrix. J Hazard Mater 272-292. doi:10.1016/j.jhazmat.2013.03.008

Dowty RA, Shaffer GP, Hester MW, Childers GW, Campo FM, Greene MC (2001) Phytoremediation of small-scale oil spills in fresh marsh environments: a mesocosm simulation. Marine Environ Res 52(3): 195-211. doi:10.1016/S0141-1136(00)00268-3

El-Ashtoukhy E-S, El-Taweel Y, Abdelwahab O, Nassef E (2013). Treatment of Petrochemical Wastewater Containing Phenolic Compounds by Electrocoagulation Using a Fixed Bed Electrochemical Reactor. Int J Electrochem Sci 8:1534 - 1550.

Eze PE, Scholz M (2008). Benzene removal with vertical - flow constructed treatment wetlands. J Chem Technol Biotechnol 83(1): 55-63. doi:10.1002/jctb.1778

Farooqi H, Basheer F, Chaudhari RJ (2008) Constructed wetland system (CWS) for wastewater treatment. In M. A. Sengupta (Ed.) *The 12th World Lake conference* pp. 1004-1009.

Fernandez-Luqueno F, Valenzuela-Encinas C, Marsch R, Martinez-Suarez C, Vazquez-Nunez E, Dendoven L (2011) Microbial communities to mitigate contamination of PAHs in soil-possibilities and challenges: a review. Environ Sci Pollut Res 12-30. doi:10.1007/s11356-010-0371-6

Ferro AM, Utah NL, Kadlec RH, Deschamp J, Wyoming C (2002) Constructed wetland system to treat wastewater at the BP Amoco former Casper refinery: Pilot scale project. *The 9th International Petroleum Envoronmental Conference.* Albuquerque, NM: IPECConsortium. Retrieved from http://ipec.utulsa.edu/Ipec/Conf2002/tech_sessions.html

Fountoulakis MS, Terzakis S, Kalogerakis N, Manios T (2009) Removal of polycyclic
aromatic hydrocarbons and linear alkylbenzene sulfonates from domestic wastewater
in pilot constructed wetlands and gravel filter. Ecol Eng 35(12): 1702-1709.
doi:10.1016/J.ECOLENG.2009.06.011

Galletti A, Verllicchi P, Rainieria E (2010) Removal and accumulation of Cu, Ni and Zn in
horizontal subsurface flow constructed wetlands: Contribution of vegetation and
filling medium. Sci Total Environ 408: 5097–5105.
doi:doi:10.1016/j.scitotenv.2010.07.045

Gillespie Jr WB, Hawkins WB, Rodgers Jr JH, Cano ML, Dorn PB (2000) Transfers and
transformations of zinc in constructed wetlands: Mitigation of a refinery effluent. Ecol
Eng 14: 279-292. doi:PII: S0925-8574(98)00113-X

Gousmi N, Sahmi A, Li H, Poncin S, Djebbar R, Bensadok K (2016) Purification and
detoxification of petroleum refinery wastewater by electrocoagulation process.
Environ Technol 1-10. doi:10.1080/09593330.2016.1150349

Grove JK, Stein OR (2005) Polar organic solvent removal in microcosm constructed
wetlands. Water Res 39(16): 4040-4050. doi:10.1016/j.watres.2005.07.023

Haberl R, Grego S, Langergraber G, Kadlec RH, Cicalini A-R, Dias SM, Novais JM, Aubert
S, Gerth A, Thomas H,Hebner, A (2003) Constructed wetlands for the treatment of
organic pollutants. J Soils Sediments 3(2): 109-124. doi:10.1007/BF02991077

Hadad HR, Maine M, Bonetto CA (2006) Macrophytes growth in a pilotconstructed wetland
for industrial wastewater treatment. Chemosphere 63: 1744-1753.
doi:10.1016/j.chemosphere.2005.09.014

Hamza UD, Mohammed IA, Sale A (2012) Potentials of bacterial isolates in bioremediation
of petroleum refinery wastewater. J Appl Phytotechnol Environ Sanitat1(3): 131-138.
Retrieved from http://www.trisanita.org/japes

Hashim M, Mukhopadhyay S, Sahu JN, Sengupta B (2011) Remediation technologies for
heavy metal contaminated groundwater. J Environ Manage 92:2355-2388.
doi:10.1016/jenvman2011.06.009

Hietala KA, Roane TM (2009) Microbial remediation of metals in soils . Adv Appl Biorem Soil Biol 17:201-220. doi:10.1007/978-3-540-89621-0_11

Hinchman RR, Negri MC, Gatliff EG (1996) Phytoremediation: using green plants to clean up contaminated soil, groundwater. *International Tropical Meeting on Nuclear and Hazardous Waste Management.* Spectrum 96, Seattle, WA: American Nuclear Society.

Ho Y, Show K, Guo X, Norli I, Abbas FA, Morad N (2012) Industrial Discharge and Their Effect to the Environment. In P. K.-Y. Show (Ed.) Industrial Waste pp. 1-32. INTECH. Retrieved January 11, 2016, from http://www.intechopen.com/books/industrial-waste/industrial-emissions-and-theireffect-

Hoffmann H, Platzer C, Winker M, von Muench E (2011) *Technology review of constructed wetlands Subsurface flow constructed wetlands for greywater and domestic wastewater treatment.* Federal Ministryfor Economic Cooperation and Development, Division Water, Energy and Transport. Eschbom: Deutsche Gesellschaft für Internationale Zusammenarbeit (GIZ) GmbH Sustainable sanitation - ecosan program.

Horner JE, Castle JW, Rodgers Jr JH, Murray Gulde C, Myers JE (2012) Design and performance of pilot-scale constructed wetland treatment systems for treating oilfield produced water from sub-saharan Africa. Water Air Soil Pollut 223: 1945-1957. doi:10.1007/s11270-011-0996-1

Hua T, Haynes R, Zhou Y-F, Boullemant A (2015) Potential for use of industrial waste materials as filter media for removal of Al, Mo, As, V and Ga from alkaline drainage in constructed wetlands - adsorption studies. Water Res 71:32-41. doi:10.1016/j.watres.2014.12.036

Huang X-D, El-Alawi Y, Gurska J, Glick BR, Greenberg BM (2005) A multi-process Phytoremediation system for decontamination of persistent total petroleum hydrocarbons (TPHs) from soils. Microchemical 81: 139-147. doi: 10.1016/j.microc.2005.01

Huddleston GM, Gillespie WB, Rodgers JH (2000) Using constructed wetlands to treat biochemical oxygen demand and ammonia associated with a refinery effluent. Ecotoxicol Environ Safety 45:188-193. doi:10.1006/eesa.1999.1852

Idise OE, Ameh JB, Yakubu SE, Okuofu CA (2010) Biodegradation of a refinery effluent treated with organic fertilizer by modified strains of *Bacillus cereus* and *Pseudomonas aeruginosa*. Afric J Biotechnol 9(22): 3298-3302. Retrieved from http://www.academicjournals.org/AJB

Ijah UJ, Antai SP (2003) The potential use of chicken drop microorganisms for oil spillage remediation. The Environmentalist 23(1):89-95. doi:10.1023/A:1022947727324

Imfeld G, Braeckevelt M, Kuschk P, Richnow HH (2009) Monitoring and assessing processes of organic chemicals removal in constructed wetlands. Chemosphere 74(3): 349-362. doi:10.1016/j.chemosphere.2008.09.062

Jadia C, Fulekar M (2009) Phytoremediation of heavy metals: Recent techniques. Afric J Biotechnol 8(6):921-928.

Jayaweera MW, Kasturiarachchi JC, Kularatne RK, Wijeyekoon SL (2008). Contribution of water hyacinth (Eichhornia crassipes (Mart.) Solms) grown under different nutrient conditions to Fe-removal mechanisms in constructed wetlands. J Environ Manage 87:450-460. doi:10.1016/j.jenvman.2007.01.013

Jechalke S, Vogt C, Reiche N, Franchini AG, Borsdorf H, Neu TR, Richnow HH (2010) Aerated treatment pond technology with biofilm promoting mats for the bioremediation of benzene, MTBE and ammonium contaminated groundwater. Water Res 44:1785-1796. doi:10.1016/j.watres.2009.12.002

Jeon BY, Jung L, Park DH (2011) Mineralization of petroleum contaminated wastewater by co-culture of petroleum-degrading bacterial community and biosurfactant-producing bacterium. Environ Protect 2(7): 895-902. doi:10.4236/jep.2011.27102

Ji GD, Sun TH, Oixing Z, Xin S, Shijun C, Peijun L (2002) Subsurface flow wetland for treating heavy oil-produced water of the Liohe Oilfield in China. Ecol Eng 18:459-465.

Ji G, Sun T, Ni J (2007) Surface flow constructed wetland for heavy oil-produced water treatment. Bioresour Technol 98(2):436-441. doi:10.1016/j.biortech.2006.01.017

Kadlec RH, Wallace SD (2009) Treatment Wetlands. (2nd ed.) New York, United States of American: CRC Press Taylor & Francis Group. http://www.taylorandfrancis.com

Kanagy LE, Johnson BM, Castle JW, Rodgers Jr JH (2008) Design and performance of a pilot-scale constructed wetland treatment system for natural gas storage produced water. Bioresour Technol, 99:1877–1885. doi:10.1016/j.biortech.2007.03.059

Khan S, Ahmad I, Shah MT, Rehman S, Khaliq A (2009) Use of constructed wetland for the removal of heavy metals from industrial wastewater. J Environ Manage 50(11): 3451–3457. doi:10.1016/j.jenvman.2009.05.026

Kivaisi, A. K. (2001) The potential for constructed wetlands for wastewater treatment and reuse in developing countries; a review. Ecol Eng 16: 545-560 . PII: S0925-854(00)00113-0

Knight R, Kadlec R, Ohlendorf HM (1999) The Use of Treatment Wetlands for Petroleum Industry Effluents. Environ Sci Technol 33(7): 973-980. doi:10.1021/es980740w

Kröpfelova L, Vymazal J, Svehla, J, Š tichova' J (2009) Removal of trace elements in three horizontal sub-surface flow constructed wetlands in the Czech Republic. Environ Pollut 157: 1186-1194. doi:10.1016/j.envpol.2008.12.003

LeDuc DL, Terry N (2005) Phytoremediation of toxic trace elements in soil and water. J Ind Microbiol Biotechnol 32:514-520. doi:10.1007/s10295-005-0227-0

Lee B-H, Scholz M (2007) What is the role of *Phragmites australis* in experimental constructed wetland filters treating urban runoff? Ecol Eng 29:87-95. doi:10.1016/j.ecoleng.2006.08.001

Lin Q, Mendelssohn IA (2009). Potential of restoration and phytoremediation with Juncus roemerianus for diesel-contaminated coastal wetlands. Ecol Eng 35(1): 85-91. http://dx.doi.org/10.1016/j.ecoleng.2008.09.010

Liu D, Ge Y, CJ, Peng C, Gu B, Chan GY, Wu X (2008) Constructed wetlands in China: recent developments and future challenges. Front Ecol Environ 7(5): 261-268. doi:10.1890/070110

Madera-Parra CA, Pena-Salamanca EJ, Pena M, Rousseau DP, Lens PN (2015) Phytoremediation of Landfill Leachate with *Colocasia esculenta*, *Gynerum sagittatum* and *Heliconia psittacorum* in Constructed Wetlands. Int J Phytorem, 17, 16-24. doi:10.1080/15226514.2013.828014

Maine MA, Suñe N, Haadad H, Sánchez G, Bonetto C (2007) Removal efficiency of a constructed wetland for wastewater treatment according to vegetation dominance. Chemosphere 68(6): 1105-1113. doi:10.1016/j.chemosphere.2007.01.064

Mant C, Coasta S, Williams J, Tambourgi E (2005) Studies of removal of chromium by model constructed wetlands. Braillian J of Chem Eng 22(3):381-387. http://dx.doi.org/10.1590/S0104-66322005000300007

Mazeikiene A, Rimeika M, Osikinis VV, Paskauskaite N, Brannvall E (2005) Removal of petroleum products from water using natural sorbent zeolite. Environ Eng Landscape Manage 13(4): 187-191. doi:10.1080/16486897.2005.9636870

Mazzeo DE, Levy CE, de Angelis DD, Marin-Morales MA (2010) BTEX biodegradation by bacteria from effluents of petroleum refinery. Sci the Total Environ 408:4334 - 4340. doi:10.1016/j.scitotenv.2010.07.004

McIntosh P, Schulthess CP, Kuzovkina YA, Guillard K (2017) Bioremediation and phytoremediation of total petroleum hydrocarbons (TPH) under various conditions. Int J Phytorem 19(8): 755-764. http://dx.doi.org/10.1080/15226514.2017.1284753

Melián JH, Rodríguez AM, Arãna, J, Díaz OG, Henríquez JG (2010) Hybrid constructed wetlands for wastewater treatment and reuse in the Canary Islands. Ecol Eng 36: 891–899. doi:10.1016/j.ecoleng.2010.03.009

Mena J, Rodriguez L, Numez J, Fermandez FJ, Villasenor J (2008) Design of horizontal and vertical subsurface flow constructed wetlands treating industrial wastewater. Water Pollut 555-557). doi:10.2495/WP080551

Merkl N, Schultze-Kraft R, Arias M (2006) Effect of the tropical grass *Brachiaria brizantha* (Hochst. ex A. Rich.) *Stapf* on microbial population and activity in petroleum-contaminated soil. Microbiol Res 80-91. doi:10.1016/j.micres.2005.06.005

Mohammed Y, Aliyu A, Audu P (2013) Concentration of Oxirane and Naphthalene in selected petroleum based industrial effluents in Kaduna metropolis, Nigeria. Direct Res JChem 1-5. http://directresearchjournals.org

Moneke A, Nwangwu C (2011) Studies on the bioutilization of some petroleum hydrocarbons by single and mixed cultures of some bacterial species. Afric J

Microbiol Res 5(12): 1457 - 1466. Retrieved from http://www.academicjournals.org/ajmr

Moreno C, Farahbakhazad N, Morrison GM (2002) Ammonia removal from oil refinery effluent in vertical upflow macrophytes column systems. Water Air Soil Pollut 135:237-247.

Mulkerrins D, Dobson A, Colleran E (2004) Parameters affecting biological phosphate removal from wastewaters. Environ Int 30:249– 259. doi:10.1016/S0160-4120(03)00177-6

Murray-Gulde C, Heatley JE, Karanfil T, Rodgers Jr JR, Myers JE (2003) Performance of a hybrid reverse osmosis-constructed wetland treatment system for brackish oil field produced water. Water Res 37(3): 705-713. doi:PII:S0043-1354(02)00353-6

Mustapha HI, Rousseau D, van Bruggen J.J.A, Lens PNL (2011) Treatment performance of horizontal subsurface flow constructed wetlands treating inorganic pollutants in simulated refinery effluent. *2nd Biennial Engineering Conference.* Minna: School of Engineering and Engineering Technology, Federal University of Technology, Minna.

Mustapha HI, van Bruggen JJA, Lens PNL (2017) Fate of heavy metals in vertical subsurface flow constructed wetlands treating secondary treated petroleum refinery wastewater in Kaduna, Nigeria. Int J Phytorem 1-10. doi:10.1080/15226514.2017.1337062

Mustapha HI, van Bruggen JJA, Lens PNL (2015) Vertical subsurface flow coonstructed wetlands for polishing secondary Kaduna refinery wastewater in Nigeria. Ecol Eng 85:588-595. doi:10.1016/j.ecoleng.2015.09.060

Mustapha HI, van Bruggen JJA., Lens PNL (2013) Preminary studies on the application of constructed wetlands for a treatment of refinery effluent in Nigeria: A mesocosm scale study. *2013 International Engineering Conference, Exhibition and Annual General Meeting.* Abuja: Nigerian Society of Engineers.

Nacheva PM (2011) Water Management in the petroleum refining industry. in D. M. Jha (Ed.), *Water Conservation* pp. 105-128. InTech. Retrieved Febuary 13, 2016, from http://www.intechopen.com/books/water-conservation/water-management-in-the-petroleum-refining-industry

Nakata C, Qualizza C, MacKinnon M, Renault S (2011) Growth and physiological responses of *Triticum aestivum* and *Deschampsia caespitosa* exposed to petroleum coke. Water Air Soil Pollut 216(1): 59-72. doi:10.1007/s11270-010-0514-x

Noori A, Maivan HZ, Alaie E, Newman L (2015) *Leucanthemum vulgare Lam.* crude oil phytoremediation. Int J Phytorem, *00*(00), 00. Retrieved from http://dx.doi.org/10.1080/15226514.2015.1045122.

Nwanyanwu CE, Abu GO (2010) In vitro effects of petroleum refinery wastewater on dehydrogenase activity in marine bacterial strains. *Ambi-Agua, Taubate,* 5(2): 21-29. doi:10.4136/ambi-agua.133

Ogunfowokan AO, Asubiojo OI, FatokiO (2003) Isolation and determination of polycyclic aromatic hydrocarbons in surface runoff and sediments. Water Air Soil Pollut 147(1): 245-261. doi:10.1023/A:1024573211382

Ojumu TV, Beelo OO, Solomon BO (2005) Evaluation of microbial Systems for Bioremediation of Petroleum Refinery Effluents in Nigeria. Afric J Biotechnolo http://dx.doi.org/10.4314%2Fajb.v4i1.15048

O'Sullivan AD, Moran BM, Otte ML (2004) Accumulation and fate of contaminants (Zn, Pb, Fe and S) in substrates of wetlands constructed for treating mine wastewater. Water Air Soil Pollut 30: 1-20.

Otokunefor T, Obiukwu C (2010) Efficacy of inorganic nutrients in bioremediation of a refinery effluent. Scientia Africana 9(1):111-125.

Otokunefor VT, Obiukwu C (2005) Impact of Refinery Effluent on the Physicochemical Properties of a Waterbody in the Niger Delta. Appl Ecol Environ Res 3(1): 61-72. Retrieved from http://www.ecology.kee.hu

Papaevangelou VA, Gikas GD, Tsihrintzis VA (2017) Chromium removal from wastewater using HSF and VF pilot-scale constructed wetlands: Overall performance, and fate and distribution of this element within the removal environment . Chemosphere 168: 716-730. Retrieved from http://dx.doi.org/10.1016/j.chemosphre.2016.11.002

Prasad MN (2003) Phytoremediation of metal-polluted ecosystems: hype for commercialization. Russian J Plant Physiol 50(5): 686-701.

Qasaimeh A, AlSharie H, Masoud T (2015) A Review on Constructed Wetlands Components and Heavy Metal Removal from Wastewater. J Environ Protect 6:710-718. doi:doi.org/10.4236/jep.2015.67064

Rani SH, Md. Din MF, Yusof BM, Chelliapan S (2011) Overview of Subsurface Constructed Wetlands Application in Tropical Climates. Universal J Environ Res Technol 1(2):103-114. Retrieved from www.environmentaljournal.org

Ravanchi MT, Kaghazchi T, Kargari A (2009) Application of membrane seperation processes in petrochemical industry: a review. Desalination 235(1-3): 199-244. doi:10.1016/j.desal.2007.10.043

Reddy KR, D' Angelo EM (1997) Biogeochemical indicators to evaluate pollutant removal efficiency in constructed wetlands. Water Sci Technol 35(5):1-10. Retrieved from http://dx.doi.org/10.1016/S0273-1223(97)

Rezaee A, Hossini H, Masoumbeigi H, Soltani RD. (2011) Simultaneous Removal of Hexavalent Chromium and Nitrate from Wastewater using Electrocoagulation Method. Int J Environ Sci Develop, 2(4): 294-298.

Rousseau DP, Lesage E. Story A, Vanrolleghen PA, De Pauw N (2008) Constructed wetlands for water reclamation. Desalination 128:181-189. doi:10.1016/j.desal.2006.09.034

Saeed T, Guangzhi S (2012) A review on nitrogen and organics removal mechanisms in subsurface flow constructeds: Dependency on environmental parameters, operating conditions and supporting media. J Environ Manage 112:), 429-448. (http://dx.doi.org/10.1016/j.jenvman.2012.08.011

Saien J (2010) Treatment of the refinery wastewater by nano particles of TiO2. *USA Patent No. Pub. No.: US 20100200515 A1.* Retrieved February 2011

Saien J, Shahrezaei F (2012) Organic pollutants removal frompetroleum refinerywastewater with nanotitania photocatalyst and UV light emission. Int J Photoenergy 2012: 1-5. doi:10.1155/2012/703074

Scholz M (2003) Performance predictions of mature experimental constructed wetlands which treat urban water receiving high loads of lead and copper. Water Res 37(6):1270-1277. doi:10.1016/S0043-1354(02)00373-1

Schröder P, Navarro-Aviñó J, Azaizeh H, Azaizeh H, Goldhirsh AG, DiGregorio S, Komives T, Langergraber G, Lenz A, Maestri E, Memon AR, Ranalli A, Sebastiani L, Smrcek S, Vanek T, Vuilleumier S, Wissing, F (2007) Using phytoremediation technologies to upgrade waste water treatment in Europe. Environ Sci Pollut Res - Int 14(7):490-497. doi:10.1065/espr2006.12.373

Scullion J (2006) Remediating polluted solis. Naturwissenschaften 93(2): 51-65. doi:10.1007/s00114-005-0079-5

Seeger EM, Kuschk P, Fazekas H, Grathwohl P, Kaestner M (2011) Bioremediation of benzene-, MTBE- and ammonia-contaminated groundwater with pilot-scale constructed wetlands. Environ Pollut 199:, 3769-3776. (doi:10.1016/j.envpol.2011.07.019)

Sepahi AA, Golpasha ID, Emami A, Nakhoda MA (2008) Isolation and Characterization of Crude Oil Degrading *Bacillus Spp*. Iran J Environ Health Sci Eng 150-152.

Shelef O, Gross A, Rachmilevitch S (2013) Role of plants in a constructed wetland: Current and New Perspectives. Water 5:405-419. doi:10.3390/w5020405

Shpiner R, Liu G, Stuckey DC (2009b) Treatment of oilfield produced water by waste stabilization ponds: biodegradation of petroleum-derived materials. Bioresourc Technol 100:6229-6235. doi:10.1016/j.biortech.2009.07.005

Shpiner R, Vathi S, Stuckey DC (2009a) Treatment of "produced water" by waste stabilization ponds: removal of heavy metals. Water Res 43:4258-4268. doi:10.1016/j.watres.2009.06.004

Šíma J, Krejsa J, Svoboda L (2015) Removal of mercury from wastewater using a constructed wetland. Croat Chem Acta 88(2):165-169. doi:org/10.5562/cca2529

Simi A (2000) Water quality assessmentof a surface flow constructed wetland treating oil refinery wastewater. In K. R. Reddy (Ed.), *7th International Conference on Wetlands System for Water Pollution Control. 3*, pp. 1295-1304. Lake Buena Vista, Boca Raton, Florida, USA: IWA. Retrieved January 17, 2011

Song X, Yan D, Liu Z, Chen Y, Lu S, Wang D (2011) Performance of laboratory-scale constructed wetlands coupled with micro-electric field for heavy metal-contaminating

wastewater treatment. Ecol Eng 37(12):2061 - 2065.
doi::10.1016/j.ecoleng.2011.08.019

Spacil M, Rodgers J, Castle J, Murray-Gulde C, Myers JE (2011) Treatment of Selenium in
Simulated Refinery Effluent Using a Pilot-Scale Constructed Wetland Treatment
System. Water Air Soil Pollut 221(1): 301-312. doi:10.1007/s11270-011-0791-z

Stefanakis AI, Seeger E, Dorer C, Sinke A, Thullner M (2016) Performance of pilot-scale
horizontal subsurface flow constructedwetlands treating groundwater contaminated
with phenols andpetroleum derivatives. Ecol Eng 95:514-526.
doi:10.1016/j.ecoleng.2016.06.105

Stottmeister U, Wießner A, Kuschk P, Kappelmeyer U, Kästner M, Bederski O, Müller R A,
Moormann, H. (2003) Effects of plants and microorganisms in constructed wetlands
for wastewater treatment. Biotechnol Adv 22: 93-117.
doi:10.1016/j.biotechadv.2003.08.010

Tam NF, Wong YS (2008) Effectiveness of bacterial inoculum and mangrove plants on
remediation of sediment contaminated with polycyclic aromatic hydrocarbons. Mater
Pollut Bulletin 57:716-726. doi:10.1016/j.marpolbul.2008.02.029

Taneva N (2012) Removal of ammonium and phosphates from aqueous solutions by
activated and modified Bulgarian clinoptilolite. J Chem Eng Mater Sci 3(5):79-85.
doi:10.5897/JCEMS11.028

Tanner CC (2001) Plants as ecosystem engineers in subsurface-flow treatment wetlands.
Water Sci Tehnol 44(11-12): 9-17.

Taylor CR, Hook PB, Stein OR, Zabinski CA (2011) Seasonal effects of 19 plant species on
COD removal in subsurface wetland microcosms. Ecol Eng 37(5): 703-710.
doi:10.1016/j.ecoleng.2010.05.007

Tirkey A, Shrivastava P, Saxena A (2012) Bioaccumulation of heavy metals in different
components of two Lakes Ecosystem. Current World Environ 7(2): 293-297.

Tromp K, Lima AT, Barendregt VT (2012) Retention of heavy metals and poly-aromatic
hydrocarbons from road water in a constructed wetland and the effect of de-icing.
Hazard Mater 203-204:290-298. doi:10.1016/j.jhazmat.2011.12.024

Uzoekwe SA, Oghosanine FA (2011) The effect of refinery and petrochemical effluent on water quality of Ubeji Creek Waar, Southern Nigeria. *Ethopian* J Environ Studies Manage *4*(2): 107-116. Retrieved from http://dx.doi.org/10.4314/ejesm.v4i2.12

Valderrama LT, Del Campo CM, Rodriguez CM, de-Bashan LE, Bashan Y (2002) Treatment ofrecalcitrant wastewater from ethanol and citric acid production using microalga *Chlorella* vulgaris and the macrophyte *Lemna minuscula*. Water Res 36(17):4185-4192. doi:PII: S0043-1354(02)00143-4

van Afferden M, Rahman KZ, Mosig P, De Biase C, Thullner M, Oswald SE, Müller RA (2011) Remediation of groundwater contaminated with MTBE and benzene: The potential of vertical-flow soil filter systems. Water Res 45:5063-5074. doi:0.1016/j.watres.2011.07.010

Veenstra J, Mohr K, Sanders DA (1998) Refinery wastewater management using multiple-angle water seperators. *The International Petroleum Environment Conference.* Albuquerque, New Mexico.

Veil JA, Puder MG, Elcock D, Redweil Jr RJ (2004) A white paper describing produced water from production of crude oil, Natural gas and coal bed methane. Department of Energy, National Energy Technology. US Department of Energy, National Energy Technology under contract W-31-109-Eng-38. doi:10.2172/821666

Vymazal J (2005) Horizontal sub-surface flow and hybrid constructed wetlands systems for wastewater treatment. Ecol Eng 25: 478-490. doi:10.1016/j.ecoleng.2005.07.010

Vymazal J (2007) Removal of nutrients in various types of constructed wetlands. Sci the Total Environ 380: 48-65. doi:10.1016/j.scitotenv.2006.09.014

Vymazal J (2010) Constructed Wetlands for Wastewater Treatment. Water 2(3):530-549. doi:10.3390/w2030530

Vymazal J (2011) Plants used in constructed wetlands with horizontal subsurface flow: a review. Hydrobiologia 674:133-156. doi:10.1007/s10750-011-0738-9

Vymazal J (2013) Emergent plants used in free water surface constructed wetlands: A review. Ecol Eng 61P: 582-592. doi:10.1016/j.ecoleng.2013.06.023

Vymazal J, Břězinová T (2016) Accumulation of heavy metals in aboveground biomass of *Phragmites australis* in horizontal flow constructed wetlands for wastewater treatment: A review. Chem EngJ 290:232-242. Retrieved from http://dx.doi.org/10.1016/j.cej.2015.12.108

Wake H (2005). Oil refineries: a review of their ecological impacts on the aquatic environment. Estuarine Coastal Shelf Sci 62: 131-140. doi:10.1016/j.ecss.2004.08.013

Wallace SD (2001) On-site remediation of petroleum contact wastes using surface flow wetlands. *2nd International Conference on Wetlands and Remediation.* Burlington.

Wallace S, Davis BM (2009) Engineered wetland design and applications for on-site bioremediation of PHC groundwater and wastewater. Society of Petroleum Engineers 4(1). doi:10.2118/111515-PA

Wallace S, Kadlec R (2005). BTEX degradation in a cold-climate wetland system. Water Sci Technol 51(9):165-171.

Wallace S, Schmidt M, Larson E (2011) Long term hydrocarbon removal using treatment wetlands. *SPE Annual Technical Conference and Exhibition* pp. 1-10. Denver, Colorado: Society of Petroleum Engineers (SPE) International.

Weis JS, Weis P (2004) Review. Metal uptake, transport and release by wetland plants: implications for phytoremediation and restoration. Environ Int 30:685-700. doi:10.1016/j.envint.2003.11.002

Wenzel WW (2009) Rhizosphere processes and management in plant-assisted bioremediation (phytoremediation) of soils. Plant Soil 321(1): 385-408. doi:10.1007/s11104-008-9686-1

Weyens N, van der Lelie D, Taghavi S, Vangronsveld J (2009) Phytoremediation: plant-endophyte partnerships take the challenge. Current Opinion Biotechnol 20(2): 248-254. doi:10.1016/j.copbio.2009.02.012

White SA (2013) Wetland technologies for nursery and greenhouse compliance with nutrient regulations. Horticultural Sci 48(9):1103-1108.

Wuyep PA, Chuma AG, Awodi S, Nok AJ (2007) Biosorption of Cr, Mn, Fe, Ni, Cu and Pb
metals from petroleum refinery effluent by *calcium alginate* immobilized *mycelia* of
Polyporus squmosus. Sci Res Essay 2(7):217-221.

Xia P, Ke H, Deng Z, Tang P (2003) Effectiveness of Constructed Wetlands for Oil-Refined
Wastewater Purification. In P. N. Truong, and X. Hanping (Ed.), *the third
International Conference on Vetiver and Exhibition : vetiver and water : an eco-
technology for water quality improvement, land stabilization, and environmental
enhancement . 22*, pp. 649-654. Guangzhou: Guangdong Academy of Agricultural
Sciences China Agriculture Press. Retrieved from
http://trove.nla.gov.au/work/37498028?versionId=48942993

Yadav AK, Abbassi R, Kumar N, Satya S, Sreekrishnan T (2012) The removal of heavy
metals in wetland microcosms: Effects of bed, plant species, and metal mobility.
Chem Eng J 211-212:501-507. doi:10.1016/j.cej.2012.09.039

Yang B, Lan C, Yang C, Liao W, Chang H, Shu W (2006) Long-term efficiency and stability
of wetlands for treating wastewater of a lead/zinc mine and the concurrent ecosystem
development. Environ Pollut 143: 466-512. doi:10.1016/j.envpol.2005.11.045

Yu L, Han M, He F (2013) A review of treating oily wastewater. Arabian J Chem 00 - 00.
Retrieved December 9, 2016, from http://dx.doi.org/10.1016/j.arabjc.2013.07.020

Zhang C-B, Liu W-L, Wang J, Ge Y, Ge Y, Chang SX, Chang J (2011) Effects of monocot
and dicot types and species richness in mesocosm constructed wetlands on removal of
pollutants from wastewater. Bioresour Technol 102:10260-10265.
doi:10.1016/j.biortech.2011.08.081

Zhang J-E, Liu J-L, Liu Ounyang Y, Liao B-W, Zhao B-L (2010) Removal of nutrients and
heavy metals from wastewater with mangrove *Sonneratia apetala Buch-Ham*. Ecol
Eng 36(10): 807-812. doi:10.1016/j.ecoleng.2010.02.008

Zhao H, Li G (2011) Application of fibrous coalescer un the treatment of oily wastewater.
Procedia Environ Sci 10:158-162. doi:10.1016/j.proenv.2011.09.028

Zheng C, Zhao L, Zhou X, Fu Z, Li A (2013) Treatment Technologies for Organic
Wastewater. In Water Treatment pp. 249 - 286. INTECH. Retrieved from
http://dx.doi.org/10.5772/52665

Ch. 3. Characterization of secondary treated refinery effluent

The main part of this chapter has been published as part of this research paper as:

Mustapha, H. I., van Bruggen, J. J., Lens, P. N. L., 2015. Vertical subsurface flow coonstructed wetlands for polishing secondary Kaduna refinery wastewater in Nigeria. Ecol. Eng. 85, 588-595. doi:10.1016/j.ecoleng.2015.09.060

Abstract

The heterogeneity of refinery effluents are characterized by the presence of large quantities of crude oil products, polyaromatic hydrocarbons (PAHs), phenols (creosols and xylenols), metals and their derivatives, ammonia, suspended solids, and sulphides. Composite samples were collected once every month from the final discharge channel of the Kaduna Refining and Petrochemical Company (KRPC) for characterization in order to obtain baseline information on the quality of effluent discharged into the environment. The measured parameters were pH, temperature, electrical conductivity, dissolved oxygen (DO), turbidity, total dissolved solids (TDS), total suspended solids (TSS), total solids (TS), biological oxygen demand (BOD$_5$), chemical oxygen demand (COD), nitrate, ammonium-Nitrogen, phosphate, potassium, sulphate, some metals (Cadmium (Cd), chromium (Cr), copper (Cu), zinc (Zn), lead (Pb) and Iron (Fe)), oil & grease and phenol using standard methods. Based on the Federal Environmental Protection Agency (FEPA) of Nigeria, European Union (EU) guidelines and World Health Organization (WHO) standards, the treated secondary refinery effluent contained high levels of BOD$_5$, COD, TSS, turbidity and some metals (Cd, Cr, and Fe). This information is intended to serve as a baseline information for the quality of effluent released into the environment. The challenges of KRPC is meeting the compliance (effluent discharge) limits as well as cost-effective effluent treatment. Hence, constructed wetlands is a viable alternative that are ecologically sustainable, economically affordable, and are effective technology that can address many of the water-management problems faced by the petroleum industry.

Keywords: Characterization; Refinery effluent; Contaminants; Quality, discharge limits

3.1   Introduction

Water is the most valuable resource of the world alongside air and soil. It represents a vital resource for a variety of human activities and also provides a living environment for an array of aquatic organisms (Adeyemo, 2003). However, the global fresh water available, particularly in developing countries, is deteriorating due to pollution and this is intensifying its shortage (Aslam et al., 2010; Kivaisi, 2001). Indeed, contaminated water is unsafe and uneconomical for domestic, agricultural and industrial purposes as well as for the environment.

Nigeria is endowed with substantial resources of water body which is estimated at 900 Km^2 (Ekiye and Zejiao, 2010), and this represent only 0.1% of its total land mass (Taiwo et al., 2012). In developing countries, rivers are used as the end point of hazardous waste discharge by industries (Jadia and Fulekar, 2009; Sikder et al., 2013). The release of untreated or partially treated effluents into water body is one of the main sources of environmental pollution, which can adversely affect its quality (Wake, 2005), municipalities and the people that live around these rivers (Kanu and Achi, 2011). Furthermore, these affected people depend heavily on these water sources of doubtful quality for domestic uses, fishery, transportation, irrigation, and recreation (Ekiye and Zejiao, 2010; Igbinosa and Okoh, 2009). Other sources of discharge of contaminants into rivers include land runoff and atmospheric fallout (dust and particulate matter) (Wake, 2005). Thus, ensuring that water body is protected from contamination is very vital to both the economy and health of the people.

3.1.1 Petroleum effluents

Petroleum is an essential instrument in the economy of Nigeria. Refining crude to finished products requires very large quantities of water. The water use is such that about 56%, 16% and 19% of the water is used in cooling systems, boiling systems, and production processes while the remaining 9% is used in auxiliary operations (Nacheva, 2011). However, petroleum refineries also generate large volume of effluent in the course of refining crude oil (Mustapha et al., 2015; Atubi, 2011). Petroleum effluents are also generated as a result of storage and distribution. These effluents are characterized by the presence of large quantities of crude oil products, poly aromatic hydrocarbons (PAHs), phenols (creosols and xylenols), metals and their derivatives, ammonia, suspended solids, and sulphides (Mustapha et al., 2015; Diya'uddeen et al., 2011; Ali et al., 2005; Otokunefor and Obiukwu, 2005). The potential toxicities of these contaminants have been reported by a number of researchers (Mustapha et al., 2015; Diya'uddeen et al, 2011; Idise et al., 2010; Mahre et al., 2007; Otokunefor and Obiukwu, 2005; Wake, 2005; Yusuff and Sonibare, 2004). For example, ammonia is harmful to fish or other aquatic organisms at free (un-ionized) concentration of 10-50 µg/l, and higher pH and sulphide in the effluent are of environmental concern, because they can lead to poor air quality of an area if not properly taken care of, thus, becoming threat to human, vegetation and materials (Yusuff and Sonibare, 2004). Phenolic compounds pose a significant threat to the environment due to their extreme toxicity, stability, bioaccumulation, and ability to remain in the environment for long period (Diya'uddeen et al., 2011). Heavy metals such as cadmium, chromium, lead, iron and zinc respectively, can adversely impact human health and aquatic ecosystems by resulting in kidney and liver damages and deformation of bone structures; cause skin cancer, convulsions and lung cancer in man; chronic intoxication can lead to encephalopathy mainly in children; iron at high concentration can increase free radicals production and is responsible for degenerative diseases and ageing and elevated zinc intake can cause muscular pain and intestinal haemorrhage (Ho et al., 2012).

The quality of petroleum refinery effluents are very much depended on the type of oil being processed, the plant configuration and operation procedures (Diya'uddeen et al., 2011; Nacheva, 2011). The fate of the oil refinery effluent once it is discharged into the environment depends on the conditions and hydrodynamics of the receiving water, this effluent is inevitably diluted within the receiving water and the extent of dilution depends on the size of the recipient and where the out fall is located whether it is intertidal or sub tidal (Wake, 2005). There is need for proper monitoring of effluents released into the environment in order to safeguard our ecosystem. According to the claim of Ho et al. (2012) discharged effluents are not well treated due to lack of highly efficient and economic treatment technology. Additionally, the Environmental Technologies Action Plan (ETPA) of the European Union claims urgent action for better water quality and protection of our natural resources with high priority given to sound environmental water treatment technologies that will reduce greenhouse gases, recycle materials and provide all partner countries with affordable technologies (Schroder et al., 2007).

Well, constructed wetland treatment technology are ecologically sustainable, economically affordable, and are effective technology that can address many of the water-management problems such as recycling and reuse, reduce greenhouse gases, and can meet stringent rules by government and regulatory authorities facing petroleum-related industries (Allen, 2008).

Constructed wetlands (CWs) are used for all types of wastewater treatment around the world (Vymazal, 2008). If they are correctly built, operated, and maintained (Rousseau et al., 2008), they can effectively restore sites of a wide variety of contaminants ranging from BOD, suspended solids, nitrogen, phosphorus, heavy metals, volatile organics, semi-volatile organics, petroleum hydrocarbons, pesticides and herbicides, PAHs, chlorinated solvents, to non-chlorinated solvents in storm water or municipal, agricultural and industrial wastewaters (Marchand et al., 2010; Imfeld et al., 2009). Additionally, CW can also supply predictable water quality (Ji et al., 2007). Therefore, the objectives of this chapter are: (1). Characterization of refinery effluent in order to have a baseline information on the quality of discharged effluent and (2). To compare the results with national (Federal Environmental Protection Agency, FEPA) and international (WHO) regulatory authorities standard limits. If need be, further treatment by using constructed wetlands.

3.2   Materials and methods

3.2.1 Description of experimental study site

The description of the study site, its location, climatic conditions and production capacity were described in chapter 1. KRPC Kaduna discharges about 100, 000 m^3 of secondary treated refinery wastewater per day. The company employs a series of physical, chemical and biological treatment methods for the generated wastewater before being released into the discharge channel. The treatment methods include: oil skimming, oxidation, biodegradation, clarification, chemical oxidation, filtration and evaporation. An investigation on the quality of this secondary discharged effluent was conducted from September 2011 to December 2012 to allow the design of the constructed wetlands used in this study (Chapter 4).

3.2.2 Physical and chemical quality characterization

Samples of secondary refinery effluent were taken monthly from the refinery effluents discharge channel prior to discharge into the environment for a period of sixteen months from the month of September 2011 through December 2012. For the purpose of this study, this secondary refinery effluent will be referred to as secondary wastewater. Secondary wastewater samples were taken just below the surface at the sampling location once monthly. The containers were thoroughly rinsed three times with the wastewater before samples were taken. Samples were collected with a plastic water sampler into 2 L labelled polyethylene containers and 250 ml capacity borosilicate glass bottles. The bottles were labelled accordingly. Table 3.1 presents analytical equipment and methods used for the characterization of secondary refinery wastewater.

In-situ measurements were carried out for pH, temperature and turbidity using handheld instruments. A portable HACH conductivity meter was used for electrical conductivity and temperature and a HANNA Instrument LP 2000 turbidity meter was used for turbidity determination. The samples were then placed in an ice-chest and convened to the laboratory for the determination of biological oxygen demand (BOD_5) chemical oxygen demand (COD), total dissolved solids (TDS), total suspended solids (TSS), nitrate-N, ammonium-N, and phosphate-P. These parameters were analyzed according to the procedures described in the Standard Methods for the Examination of Water and Wastewater (APHA, 2002): Open reflux, titrimetric method for COD (Maine et al., 2009); 5-Day incubation method for BOD_5 (Maine et al., 2009); gravimetric methods for TDS, TSS and spectrophotometric analysis for phosphate, spectrophotometric analysis for nitrate-N and ammonium-N (HACH, 1997). The samples that could not be analyzed same day were refrigerated.

The heavy metal (cadmium, chromium, copper, zinc, lead and iron) contents in the discharged secondary refinery wastewater were investigated. Heavy metals were determined using atomic absorption spectrophotometer (AAS) according to the procedures described in the Standard Methods for the Examination of Water and Wastewater (APHA, 2002).

The concentrations of phenol and O&G in the discharged secondary refinery wastewater were analyzed by ASTM D 1783 for phenol and gravimetric method after solvent extraction with xylene for oil and grease determination according to the procedure described in the Standard Methods for the Examination of Water and Wastewater (APHA, 2002).

Table 3.1: Analytical equipment and methods for characterization of refinery effluents

Parameter	Symbol	Equipment/Analytical Method
pH	pH	HORIBA pH meter
Temperature, ^0C	T^0C	HORIBA pH meter
Electrical conductivity, $\mu S/cm$	Ec	HANNA Instrument EC 215
Turbidity, TNU	-	HANNA Instrument LP 2000
Dissolve Oxygen, mg/L	DO	Winkler's Modification Method
Total Suspended Solids, mg/L	TSS	Gravimetric after filtration

Total Dissolve Solids, mg/L	TDS	Gravimetric
Biological Oxygen demand, mg/L	BOD_5	5-Day Incubation
Chemical oxygen demand, mg/L	COD	Dichromate digestion method
Ammonium Nitrogen, mg/L	NH_4^+-N	HACH DR/2010 Portable logging spectrophotometer
Nitrate, mg/L	NO_3^--N	HACH DR/2010 Portable logging spectrophotometer
Phosphate, mg/L	PO_4^{3-}-P	Spectrophotometry 752 UV
Potassium , mg/L	K	Flame Emission Photometric Method
Cadmium, mg/L	Cd	Atomic Absorption Spectrophotometry
Chromium, mg/L	Cr	Atomic Absorption Spectrophotometry
Copper, mg/L	Cu	Atomic Absorption Spectrophotometry
Zinc, mg/L	Zn	Atomic Absorption Spectrophotometry
Iron, mg/L	Fe	Atomic Absorption Spectrophotometry
Lead, mg/L	Pb	Atomic Absorption Spectrophotometry
Phenol, mg/L	-	ASTM D 1783
Oil and Grease, mg/L	O&G	Gravimetric method

3.2.3 Data analysis

The rank-based nonparametric Mann-Kendall (MK) statistical test (Kendall, 1975; Mann, 1945), a spreadsheet (Makesen 1.0) developed at the Finnish Meteorological Institute (Salmi et al., 2002) was used to assess the significance of monotonic trends in time series (Samsudin et al., 2017; Zeleňáková et al., 2017; Anghileri et al., 2015; Meals et al., 2011) to detect the significant trends. The MK was used because it is a nonparametric test which does not require the data to be normally distributed and it has a low sensitivity to abrupt breaks due to inhomogeneous time series (Karmeshu, 2012). The trend-free pre-whitening method was used before applying the MK test to detect significant trends. The Mann-Kendell test statistic S is given as:

$$S = \sum_{k-1}^{n-1} \sum_{j=k+1}^{n} \text{sgn}\left(x_j - x_k\right)$$

(3.1)

where n is the length of the time series $X_i \ldots X_n$, $\mathrm{sgn}(.)$ is a sign function, X_j and X_k are values in years j and k, respectively. The expected value of S equals zero $(E[S]=0)$ for series without trend and the variance is computed as:

$$(S) = \frac{1}{18}\left[n(n-1)(2n+5) - \sum_{p-1}^{q} t_p (t_p - 1)(2t_p + 5) \right]$$

(3.2)

Here q is the number of tied groups and t_p is the number of data values in p^{th} group. The test statistic Z is then given as:

$$z = \begin{cases} \dfrac{s-1}{\sqrt{\sigma^2(s)}} & \text{if } s > 0 \\[2mm] 0 & \text{if } s = 0 \\[2mm] \dfrac{s+1}{\sqrt{\sigma^2(s)}} & \text{if } s < 0 \end{cases}$$

(3.3)

3.3   Results

3.3.1 Characterization of secondary refinery wastewater

The physicochemical quality of the secondary wastewater discharged into the discharge channel is presented in Table 3.2. The characterized secondary refinery wastewater was composed of organic and inorganic compounds including salts, suspended solids and metals. The composition varied with the production processes.

The pH varied between 6.3 ± 0.0 and 8.5 ± 0.1 (n=32) within the sixteen months of data collection. There was a significant difference (P<0.05) in the observed pH values among the different months throughout the duration of the characterization. Temperature ranged from 21.5 ± 0.7 to 31.2 ± 0.4 ^{0}C with September 2012 having the highest temperature (31.2 ± 0.4 ^{0}C).

The mean turbidity concentrations varied significantly among the months with mean effluents turbidity values ranging from 12.2 ± 0.3 to 253.0 ± 0.7 NTU. These values were higher than the allowable limit of 5.0 NTU. Turbidity was significantly different among the months throughout the period of assessment.

Total dissolved solids (TDS) comprise all the dissolved material present in a wastewater. The range of TDS (146.7 ± 0.1 to 446.0 ± 0.4 mg/L) in the secondary wastewater was within the allowable limit of the WHO. On the other hand, total suspended solids (TSS) were generally observed to be above the limit (<30 mg/L) set by the Federal Environmental Protection Agency (FEPA) of Nigeria. The mean TSS concentration in the effluent was significantly different ($P<0.05$) between the months, however, the mean concentrations for March and April, 2012 were not significantly different ($P>0.05$) from each other. The refinery wastewater had a mean BOD_5 concentration that varied between 35.2 ± 34.9 and 283.8 ± 1.0 mg/L.

The mean nitrate-N concentrations in the secondary wastewater were generally low ranging from 0.3 ± 0.1 to 3.4 ± 0.1 mg/L, while the mean ammonium concentration was between 0.4 ± 0.1 and 12.6 ± 0.1 mg/L. There was significant difference ($P<0.05$) in the mean concentration of ammonium-N within some observed months, while no significant differences ($P>0.05$) were observed in other months. Meanwhile, there was a high significant difference in the mean phosphate-P concentrations among the observed months, but the level of phosphate-P recorded in the months of March and May 2012 were not significantly different from each other at $P>0.05$. The mean phosphate-P concentration ranged from 1.3 ± 0.0 to 15.7 ± 0.0 mg/L.

3.3.2. Heavy metal concentrations in secondary refinery wastewater

The results of the secondary refinery wastewater are presented on Table 3.3. The mean heavy metal concentrations in the secondary wastewater showed that Cd, Cu, Pb and Zn were within the safe limits of effluent discharge while Cr and were above the WHO recommended limits for drinking water (Table 3.3). However, the concentrations ranged between 0.00 – 0.03 mg/L for Cd, 0.01 – 0.05 mg/L for Cu, 0.01 – 0.06 mg/L for Pb, 0.03 – 0.80 mg/L for Zn, 0.01 – 3.4 mg/L for Cr and 0.5 – 16.9 mg/L for Fe. The monthly mean concentration of the heavy metals were found to be generally in this order of increasing magnitude of Cd < Cu < Pb < Zn < Cr < Fe.

3.3.3. Special organic contaminant concentrations in secondary refinery wastewater

The phenol and oil and grease concentrations in the discharged secondary treated refinery ranged from 0.01 - 1.16 µg/L and from 0.7 - 14.2 mg/L, respectively. The results is presented in Table 3.4. The mean phenol concentrations in the discharged effluent were higher than the minimum recommended value of < 0.1 µg/L while the mean oil and grease concentrations were within the recommended limit of 10 mg/L.

3.3.4 M-K test

Table 3.5 presents the results from the trend analysis based on the Mann-Kendall nonparametric statistical test for a critical probability run at significance levels of 5 and 10% on time series data for each of the 23 parameters coupled with the Theil-Sen approach applied to the data set for the time period of 16 months. The trend analysis of 16 months study for the 23 parameters revealed that 6 parameters: DO, TDS, TS, COD, Cr and Fe (Table 3.5) showed negative significant trends, implying decrease in the concentrations of the parameters during the period under consideration. In addition, out of the remaining 17 parameters, 8 (temperature, phosphate, potassium, sulphate, cadmium, zinc, phenol and oil and grease) display insignificant positive trends (i.e., increase in concentrations over time), while the rest 9 (pH, Ec, turbidity, TSS, BOD_5, nitrate, ammonium, copper and lead) depict negative trends that are not statistically significant. The rate of change was highest in TS (-18.35), but least in phenol (0.0021). However, three parameters – cadmium, chromium and lead showed an interesting trend as they show no change. With the application of the Theil-Sen Estimator, it thus implied that there was no strong changes in the concentration of the parameters recorded over the 16 months period. However, the insignificant positive trends of temperature, phosphate, potassium, sulphate, cadmium, zinc, phenol and oil and grease have led to the decreasing concentration of DO indicating contamination of the discharged effluent. Aside the fact that increased temperature reduces the amount of DO, it also increases the rate of evapotranspiration (Adonadaga, 2014) increasing the concentrations of the contaminants in the discharged effluent. Thus, this calls for further treatment or polishing of the secondary refinery effluent.

Table 3.2: Characteristics of physio-chemical quality of secondary refinery wastewater from September 2011 to December 2012

Parameter	Sept 2011	Oct 2011	Nov 2011	Dec 2011	Jan 2012	Feb 2012	Mar 2012	Apr 2012	May 2012	Jun 2012	Jul 2012	Aug 2012	Sept 2012	Oct 2012	Nov 2012	Dec 2012	Mean	Min	Max	Recommended limits
pH	7.0	7.5	6.8	8.4	6.9	8.5	7.8	7.4	6.7	6.6	6.5	8.3	6.6	6.5	6.6	7.4	7.2	6.5	8.5	6.0-9.0
Temperature(^{0}C)	22.0	27.6	24.7	25.6	28.9	26.1	28.0	24.0	27.1	26.6	24.7	27.4	31.1	28.8	29.8	24.8	26.7	22.0	31.1	30-36
Ec (μS/cm)	1297	1548	1564	1643	1957	1113.3	1241	1206	1355	1544	1167.1	1165.7	1173.0	1270	1471	1232	1372	1113.3	1957	1000
DO (mg/L)	2.0	1.6	1.3	1.5	1.0	1.2	0.8	1.9	1.4	1.3	1.7	1.1	1.0	1.4	1.2	1.4	1.4	0.8	2.0	<0.2
Turbidity (TNU)	36.6	50.0	38.3	253.0	45.2	13.3	26.7	33.6	59.0	71.0	16.4	136.0	43.6	21.0	53.3	12.2	56.8	12.2	253.0	5
TDS (mg/L)	331.3	378.9	402.0	417.9	446.3	402.7	199.1	203.8	127.2	216.2	181.8	181.2	156.6	238.0	385.5	146.7	276.0	127.2	446.3	500-2000
TSS (mg/L)	22.7	19.8	49.3	301.8	336.3	153.6	107.0	107.2	109.8	27.8	1.1	50.2	5.1	4.0	45.0	38.6	86.2	1.1	336.3	30-50
TS (mg/L)	354.0	398.7	451.4	719.7	782.5	556.3	306.1	311.0	237.0	244.0	182.9	231.4	161.7	242.0	430.5	185.3	362.2	161.7	782.5	-
BOD (mg/L)	132.3	70.6	110.3	102.9	117.6	119.2	119.2	101.3	140.1	110.3	59.6	60.0	10.4	61.1	283.1	118.2	107.3	10.4	283.1	10-25
COD (mg/L)	235.3	278.9	273.4	362.9	234.4	291.2	208.0	260.0	340	272.0	96.0	134.6	40.2	80.6	520.8	84.3	232.7	40.2	520.8	40-60
NO$_3$-N (mg/L)	0.7	0.6	0.7	3.3	0.3	0.3	0.7	0.4	0.9	1.0	0.3	0.3	0.4	0.3	1.0	0.3	0.7	0.3	3.3	20-50
NH$_4^+$-N (mg/L)	1.7	11.8	0.5	10.9	2.1	12.5	2.9	0.7	1.2	2.5	0.3	0.4	0.6	0.4	8.2	0.8	3.6	0.3	12.5	0.1-1.0
PO$_4^{3-}$-P (mg/L)	7.0	3.9	2.2	15.6	6.4	2.8	6.3	2.9	6.3	4.8	1.3	2.4	3.2	5.0	8.1	8.3	5.4	1.3	15.6	5

Table 3.3: Characteristics of heavy metals in secondary refinery wastewater from September 2011 to December 2012

Parameter	Sept 2011	Oct 2011	Nov 2011	Dec 2011	Jan 2012	Feb 2012	Mar 2012	Apr 2012	May 2012	Jun 2012	Jul 2012	Aug 2012	Sept 2012	Oct 2012	Nov 2012	Dec 2012	Mean	Recommended limits
Cd mg/L	<0.002	<0.002	<0.002	<0.01	<0.01	<0.01	<0.01	0.00	0.01	0.03	<0.001	<0.01	<0.01	0.02	0.01	0.01	0.005	0.002-1.0
Cr (mg/L)	<0.01	3.4	2.8	0.08	ND	ND	ND	ND	ND	ND	ND	ND	ND	ND	ND	0.01	0.4	0.05
Cu(mg/L)	<0.05	<0.5	<0.05	<0.01	0.03	0.01	<0.001	0.03	0.03	0.02	<0.01	<0.01	0.04	<0.001	<0.001	0.05	0.01	1.5
Zn (mg/	0.80	<0.5	<0.05	0.10	0.28	0.13	0.03	0.11	0.26	0.20	0.09	0.10	0.06	0.15	0.27	0.12	0.17	1.0-3.0
Fe (mg/L)	8.3	16.9	5.3	1.2	1.7	0.5	1.0	1.3	1.4	5.4	0.8	1.6	1.2	0.7	2.8	0.7	3.2	20
Pb mg/L)	<0.01	<0.01	<0.01	0.02	<0.02	0.06	<0.02	0.02	0.06	<0.01	0.01	<0.001	0.06	<0.001	<0.01	0.02	0.015	0.05

Key note: ND - Not detected

Table 3.4: Characteristics of organic contaminants in secondary refinery wastewater from September 2011 to December 2012

Parameter	Sept 2011	Oct 2011	Nov 2011	Dec 2011	Jan 2012	Feb 2012	Mar 2012	Apr 2012	May 2012	Jun 2012	Jul 2012	Aug 2012	Sept 2012	Oct 2012	Nov 2012	Dec 2012	Mean	Min	Max	Recommended limits
Phenol (µg/L)	<0.1	0.10	0.10	0.10	0.10	0.12	0.26	<0.01	0.24	ND	0.12	<0.01	0.29	0.13	1.16	0.01	0.17	0.01	1.16	<0.1
O&G (mg/L)	<0.1	2.6	4.5	<0.1	<0.1	1.4	3.89	<1.0	14.2	ND	2.5	<1.0	5.3	0.7	10.1	0.9	2.9	0.7	14.2	10.0

Key note: ND - Not detected

Table 3.5. Summary of statistical significant trends of the physical, chemical and organic quality of Kaduna Refinery wastewater from a 16-month data series from September 2011 – December 2012

Parameter	Test Z	Significance	Slope
pH	-1.4857	-	-0.0381
Temperature (0C)	1.5323	-	0.2600
Ec (µS/cm)	-1.3957	-	-15.3482
DO (mg/L)	-2.3862	*	-0.0767
Turbidity (TNU)	-0.4952	-	-0.8900
TDS (mg/L)	-1.9360	+	-12.6903
TSS (mg/L)	-1.3057	-	-5.1810
TS (mg/L)	-2.4762	*	-18.6648
BOD (mg/L)	-0.4507	-	-0.9982
COD (mg/L)	-1.6658	+	-13.4873
Nitrate (mg/L)	-0.8865	-	-0.0063
Ammonia N (mg/L)	-1.3957	-	-0.1344
Phosphate (mg/L)	0.0901	-	0.0351
Potassium (mg/L)	1.5898	-	0.2027
Sulphate (mg/L)	0.9014	-	1.4714
Cadmium (mg/L)	1.2151	-	0.0000
Chromium (mg/L)	-1.9345	+	0.0000
Copper (mg/L)	-1.6167	-	-0.0024
Zinc (mg/L)	0.3158	-	0.0015
Iron (mg/L)	-1.6658	+	-0.1289
Lead (mg/L)	-0.1429	-	0.0000
Phenol (µg/L)	0.8745	-	0.0021
Oil and grease (mg/L)	0.5881	-	0.0697

* Significant at $\alpha = 0.05$, + significant at $\alpha = 0.1$ and - not significant

3.4. Discussion

3.4.1 Implication of the discharge of secondary treated refinery wastewater into the environment

The secondary refinery wastewater revealed that pH, temperature and nitrate-N were within the permissible limits. The nature of the secondary refinery wastewaters was such that it ranged between weakly acidic and weakly basic (6.3 ± 0.0 and 8.5 ± 0.1). A similar occurrence was reported for oil refinery effluents by Gulshan and Dasti (2012). The low pH of the oil refinery wastewater can induce the dissociation of iron phosphate in solution and vice versa, also the high pH can cause more carbonate and bicarbonate in water (Lawson, 2011).

Turbidity, total dissolved solids (TDS), biological oxygen demand (BOD_5), chemical oxygen demand (COD), ammonium-N and phosphate-P were above the permissible limits. High turbid water is often associated with the possibility of microbiological contamination (Igbinosa and Okoh, 2009). Also, if such effluent is discharged into water bodies, it will affect fish and aquatic life (Akan et al., 2008).

The BOD_5 and COD concentrations were above the recommended limits of 10 mg/l and 40 mg/l, respectively allowed by both the WHO and FEPA in discharged effluents (Nwanyanwu and Abu, 2010). Discharge of effluents into the environment with high levels of BOD_5 and COD imply that less oxygen is available for living organisms (Kaur et al., 2010). In addition, this may indicate toxic conditions and the presence of biologically resistant organic substances in the effluent (Yusuff and Sonibare, 2004; Mahre et al., 2007).

Nutrients are required by plants for growth; however, high concentrations of nutrients are largely responsible for eutrophication, depletion of dissolved oxygen and pollution of water bodies (Chang et al., 2010). The values of ammonium-N in the discharged effluents generally exceeded the FEPA limits of discharged effluent. In the majority of the months (9 out of 16 months), the mean phosphate-P concentrations exceeded the FEPA limits of 5 mg/l (Israel et al., 2008). This study showed that it is paramount to further polish these contaminants in the treated refinery effluent to non-hazardous levels to protect the aquatic ecosystem and people downstream of the river who use the river as source of water for domestic and agricultural purposes.

3.4.2. Heavy metal content in the secondary oil refinery wastewater

Generally, cadmium concentrations were low and the values were below the FEPA set limits of 1.0 mg/l while the few times Cr was detected in the discharged effluents, the values were above the permissible limits. Well, chromium (III) compounds are less damaging to health due to their limited absorptions by the body (<1%). However, chromium (VI) compounds are acutely poisonous and on contact with the skin, it triggers dermatitis, allergies and irritations, it is thus considered carcinogenic to humans (Ali et al., 2005). Fe is an essential nutrient for blood and skeleton, with very high concentration in the body leading to tissue damage and hyperhaemoglobularia (Ali et al., 2005). Although, the concentration of Fe was the highest, it was still within the FEPA set limit of 20 mg/l but above the WHO recommended limits of 0.1

mg/L for drinking water. The Zn concentrations were low for all the observed months but accumulations in the receiving bodies with time will be problematic to aquatic ecosystems. The concentrations of Pb were all below the recommended limits in the observed months. Of course, accumulation of lead with time, if the discharged effluent is not well treated will result into neurological damage of fetuses, abortion and other complications in children under three years old (Ali et al., 2005).

3.4.3 Organic contaminants

Oily refinery wastewater is a mixture of hydrocarbons and phenol (Ishak et al., 2012) and the discharge of such wastewater into waterbodies have detrimental impacts on the environmental and human health. Phenol is toxic to aquatic life and lead to liver, lung, kidney and vascular system infection (Ishak et al., 2012). Thus, this wastewater need to be adequately treated before being released into the environment.

3.5 Conclusions
- The Characterization of discharged secondary treated refinery wastewater from KRPC Kaduna was investigated in this study from September 2011 to December 2012. The results revealed that TSS, turbidity, BOD_5, COD, sulphate and phenol in the discharged effluents were above the recommended limits of FEPA and WHO. The treatment methods employed by KRPC cannot sufficiently remove the contaminants to desired recommended limits.
- Secondary refinery wastewater is an indisputable pollution source for water courses.
- The concentration of Cd, Cu, Zn, Fe and Pb in the characterized effluents were below the threshold limits except for Cr.
- The Mann-Kendell trend analysis showed negative significant trends for most of the measured parameters while cadmium, chromium and lead showed no change.

3.6 Acknowledgement
The authors acknowledge the management of Kaduna Refinery and Petrochemical Company, Kaduna, Nigeria for giving us the opportunity to conduct this research in their company. The Government of the Netherlands for their financial assistance (NFP-PhD CF7447/2011).

3.7 References

Adeyemo, O. (2003). Consequences of Pollution and Degradation of Nigerian Aquatic Environment on Fisheries Resources. The Environmentalist 23: 297-306.

Adonadaga, M.-G. (2014). Climate change effects and implications for wastewater treatment options in Ghana. Journal of Environment and Earth Science 4(8): 9-17.

Akan, J. C., Abdurrahman, F. I., Dimari, G. A., Ogugbuaja, V. O. (2008). Physicochemical determination of pollutants in wastewater and vegetable samples along the Jakara wastewater channel in Kano metropolis, Kano State, Nigeria. European Journal of Scientific Research 23(1):122-133. ISSN 1450-216X. http://www.eurojournals.com/ejsr.htm

Ali, N., Oniye, S. J., Balarabe, M. L., Auta, J. (2005). Concentration of Fe, Cu, Cr, Zn and Pb in Makera-Drain, Kaduna, Nigeria. ChemClass Journal 2: 69-73.

Allen, W. E. (2008). Process water treatment in canada's oil sands industry: ii. a review of emerging technologies. Journal of Environmental Engineering and Science 7: 499-524. doi:10.1139/S08-020

Anghileri, D., Pianosi, F., Soncini-Sessa, R. (2015). Trend detection in seasonal data: from hydrology to water resources. Journal of Hydrology 511:171-179. doi:10.1016/j.jhydrol.2014.01.022

APHA. (2002). Standard methods for the examination of water and wastewater (20th ed.). Baltimore, Maryland, USA, Maryland, USA: American Public Health Association.

Aslam, M. M., Malik, M., Braig, M. A. (2010). Removal of metals from refinery wastewater through vertical flow constructed wetlands. International Journal of Agriculture and Biology 12:796-798. Retrieved from http://www.fspublishers.org

Atubi, A. O. (2011). Effect of Warri refinery effluents on water quality from the Lffie River, Delta State, Nigeria. American Review of Political Economy 45-56. Retrieved June 20, 2016, from https://sites.bemidjistate.edu/arpejournal/wp-content/uploads/sites/2/2015/11/v9n1-atubi.pdf

Chang, N.-B., Xuan, Z., Daranpob, A., Wanielista, M. (2010). A Subsurface up flow wetland system for removal of nutrients and pathogens in on-site sewage treatment and disposal systems. Environmental Engineering Science 1-14. doi:10.1089/ees.2010.0087

Diya'uddeen, B. H., Wad Daud, M. A., AbdulAziz, A. R. (2011). Treatment technologies for petroluem refinery effluent: A review. Process Safety and Environmental Protection 89:95-105. doi:10.1016/j.psep.2010.11.003

Ekiye, E., Zejiao, L. (2010). Water quality monitoring in Nigeria: Case study of Nigeria's industrial cities. Journal of American Science 6(4): 22-28. Retrieved June 20, 2016, from http://www.americanscience.org/

Gulshan, A. B., Dasti, A. A. (2012). hysico - chemical nature of oil refinery effluents and it's effects on seed germination of certain plant species. ARPN Journal of Agricultural Biological Science 7(5): 342-345. ISSN 1990-6145

HACH. (1997). Water Analysis Handbook (3rd ed.). Loveland, Colorado, USA: HACH Company.

Ho, Y., Show, K., Guo, X., Norili, I., Alkarkhi Abbas, F. M., Morad, N. (2012). Inustrial discharge and their effects to the environment. In: P. K.-Y. Show (Ed.) Industrial Waste. pp. 1-33. Croatia: INTECH. Retrieved June 22, 2016, from www.intechopen.com

Idise, O. E., Ameh, J. B., Yakubu, S. F., Okuofo, C. A. (2010). Biodegradation of a refinery effluent treated with organic fertilizer by modified strains of *Bacillus cereus* and *Pseudomonas aeruginosa*. African Journal of Biotechnology 9(22): 3298-3302. doi:10.5897/AJB10.229

Igbinosa, E. O., Okoh, A. I. (2009). Impact of discharge wastewater effluents on the physico-chemical qualities of a receiving watershed in a typical rural community. Internatioal Journal of Environmental Science and Technology 6(2):175-182. Retrieved June 22, 2016

Imfeld, G., Braeckevelt, M., Kuschk, P., Richnow, H. H. (2009). Monitoring and assessing processes of organic chemicals removal in constructed wetlands. Chemosphere, 74(3), 349-362. doi:10.1016/j.chemosphere.2008.09.062

Ishak, A., Malakahmad, A., Isa, M. (2012). Refinery wastewater biological treatment: A short review. Journal of Science and Industrial Research 71:251-258.

Israel, A. U., Obot, I. B., Umoren, S. A., Mkepenie, V., Ebong, G. A. (2008). Effluents and solid waste analysis in a petrochemical company- a case study of Eleme Petrochemical Company Ltd, Port Harcourt, Nigeria. E-Journal of Chemistry 5(1): 74-80. Retrieved from http://www.e-journals.net

Jadia, C., Fulekar, M. (2009). Phytoremediation of heavy metals: Recent techniques. African Journal of Biotechnology 8(6):921-928.

Ji, G. D., Sun, T. H., Ni, J. R. (2007). Surface flow constructed wetlands for heavy oil-produced water treatment. Bioresources Technology 98:436-441. doi:10.1016/j.biortech.2006.01017

Kanu, I., Achi, O. K. (2011). Industrial effluents and their impact on water quality of receiving rivers in Nigeria. Journal of Applied Technology Sciences Research 2(4):304-311.

Karmeshu, N. (2012). Trend detection in annual temperature and precipitation using the Mann Kendall Test – A case study to assess climate change on select states in the northeastern United States. University of Pennsylvania, Department of Earth & Environmental Science. Pennsylvania: University of Pennsylvania. Retrieved from http://repository.upenn.edu/mes_capstones/47

Kaur, A., Vats, S., Rekhi, S., Bhardwaj, A., Goel, J., Tanwar, R. S., Gaur, K. K. (2010). Physico-chemical analysis of the industrial effluents and their impact on the soil microflora. Procedia Environmental Science 2: 595-599. doi:10.1016/j.proenv.2010.10.065

Kendall, M. (1975). Rank Correlation Methods. London: Charles Griffin.

Kivaisi, A. K. (2001). The potential for constructed wetlands for wastewater treatment and reuse in developing countries; a review. Ecological Engineering 16: 545-560. PII: S0925-854(00)00113-0

Kumar, N., Bauddh, K., Kumar, S., Dwivedi, N., Singha, D. P., Barman, S. (2013). Accumulation of metals in weed species grown on the soil contaminated with industrial waste and their phytoremediation potential. Ecological Engineering 61:491-495. Retrieved from http://dx.doi.org/10.1016/j.ecoleng.2013.10.004

Lawson, E. O. (2011). Physico-chemical parameters and heavy metal contents of water from the mangrove swamps of Lagos Lagoon, Lagos, Nigeria. Advance Biological Research 5(1):8-21.

Mahre, M. Y., Akpan, J. C., Moses, E. A., Ogugbuaja, V. O. (2007). Pollution indicators in River Kaduna, Nigeria. Trends Applied Science Research 2(4): 304-311.

Maine, M. A., Sune, N., Hadad, H. R., Sanchez, G., Bonetto, C. (2009). Influence of vegetation on the removal of heavy metals and nutrients in a constructed wetland. Environmental Management 90:355-363. doi:10.1016/j.jenvman.2007.10.004

Mann, H. (1945). Nonparametric tests against trend. Econometrica , 13, 245–259.

Marchand, L., Mench, M., Jacob, D. L., Otte, M. L. (2010). Metal and metalloid removal in constructed wetlands, with emphasis on the importance of plants and standardized measurements: A review. Environmental Pollution 158(12):3447-3461. doi:10.1016/j.envpol.2010.08.018

Meals, D., Spooner, J., S.A, D., Harcum, J. B. (2011). Statistical analysis for monotonic trends. 23. Tetra Tech Inc. Fairfax, VA: U.S. Environmental Protection Agency.

Mustapha, H. I., van Bruggen, J. J. A., Lens, P. N.L (2015). Vertical subsurface flow constructed wetlands for polishing secondary Kaduna refinery wastewater in Nigeria. Ecological Engineering 84: 588-595. doi:10.1016/j.ecoleng.2015.09.060

Nacheva, P. M. (2011). Water management in the petroleum refining industry. In: M. Jha and M. Jha (Ed.), Water Conservation. pp. 105-128. Mexico: INTECH. Retrieved May 12, 2013

Nwanyanwu, C. E., Abu, G. O. (2010). In vitro effects of petroleum refinery wastewater on dehydrogenase activity in marine bacterial strains. Ambi-Agua, Taubatè 5(2): 21-29. doi:10.4136/ambi-agua.133

Otokunefor, V. T., Obiukwu, C. (2005). Impact of refinery effluent on the physicochemical properties of a waterbody in the Niger Delta. Applied Ecology and Environmental Research, 3(1): 61-72. Retrieved from http://www.ecology.kee.hu

Rousseau, D. P., Lesage, E., Story, A., Vanrolleghen, P. A., De Pauw, N. (2008). Constructed Wetlands for Water Reclamation. Desalination 128(1-3):181-189. doi:10.1016/j.desal.2006.09.034

Salmi, T., Määttä, A., Anttila, P., Ruoho-Airola, T., Amnell, T. (2002). Detecting trends of annual values of atmospheric pollutants by the Mann-Kendall Test and Sen's slope estimates. . Helsinki, Finland: Publications on Air Quality No. 31. ISBN 951-697-563-1. ISSN 1456-789X.

Samsudin, M. S., Khalit, S. I., Juahir, H., Mohd Nasir, M. F., Kamarudin, M. K., Lananan, F. (2017). Application of Mann-Kendall in Analyzing Water Quality Data Trend at Perlis River, Malaysia. International Journal on Advanced Science Engineering Information Technology 7(1):78-85.

Schröder, P., Navarro-Aviñó, J., Azaizeh, H., Azaizeh, H., Goldhirsh, A. G., DiGregorio, S., . . . Wissing, F. (2007). Using phytoremediation technologies to upgrade waste water treatment in Europe. Environmental Science and Pollution Research - International 14(7): 490-497. doi:10.1065/espr2006.12.373

Sikder, M. T., Kihara, Y., Yasuda, M., Mihara, Y., Tanaka, S., Odgerel, D., Kurasaki, M. (2013). River water pollution in developed and developing countries: judge and assessment of physicochemical and characteristics and selected dissolved metal concentration. CLEAN-Soil Air Water 41(1): 60-68. doi:10.1002/clen.201100320

Taiwo, A. M., Olujimi, O., Bamgbose, O., Arowolo, T. (2012). Surface water quality monitoring in Nigeria: Situational analysis and future management strategy, water quality monitoring and assessment. In D. Voudouris (Ed.), Water quality monitoring and assessment. pp. 301-320. InTech. Retrieved June 20, 2016, from www.intechopen.com

Vymazal, J. (2008). Constructed Wetlands for Wastewater Treatment: A Review. In: M. A. Sengupa (Ed.), Proceedings of Taal2007: The 12th World Lake Conference, pp. 965-980.

Wake, H. (2005). Oil refineries: a review of their ecological impacts on the aquatic environment. Estuarine Coastal and Shelf Science 62: 131-140. doi:10.1016/j.ecss.2004.08.013

Yusuff, R. O., Sonibare, J. A. (2004). Characterization of textile industries' effluents in Kaduna, Nigeria and pollution implications. Global Nest: the International Journal 212-221.

Zeleňáková, M., Vido, J., Portela, M. M., Purcz, P., Blištán, P., Hlavatá, H. H. (2017). Precipitation trends over Slovakia in the period 1981–2013. Water 9:1-20. doi:10.3390/w9120922

Ch. 4. Vertical subsurface flow constructed wetlands for polishing secondary Kaduna refinery wastewater in Nigeria

This chapter has been presented and published as:

Hassana Ibrahim Mustapha., Bruggen van J. J. A., P. N. L. Lens., 2013. Preliminary studies on the application of constructed wetlands for treatment of refinery effluent in Nigeria: A mesocosm scale study. In: Proceedings of the 2013 Nigerian Society of Engineers' International Engineering Conference, Exhibition and Annual General Meeting, Abuja, Nigeria (9 - 13th December 2013).

Mustapha, H. I., van Bruggen, J., Lens, P. L., 2015. Vertical subsurface flow coonstructed wetlands for polishing secondary Kaduna refinery wastewater in Nigeria. Ecol. Eng. 85, 588-595. doi:10.1016/j.ecoleng.2015.09.060

Abstract

Secondary wastewater discharged by the Kaduna refining and petrochemical company in Kaduna (Nigeria) was characterized and treated in six vertical subsurface flow constructed wetlands (VSF CW) under field conditions. The secondary refinery wastewater had high levels of BOD_5 (106 ± 58.9 mg/L), COD (232 ± 121.2 mg/L), TSS (86.1 ± 99.7 mg/L), TDS (278.8 ± 112.7 mg/L) and a turbidity of 56.8 ± 59.2 NTU. *Cyperus alternifolius* and *Cynodon dactylon* (L.) Pers. were planted in four VSF CWs and the last two VSF CWs served as the unplanted control. The VSF CWs were operated to study if they can further reduce the concentrations of the contaminants of interest. Good reductions in the concentration of the contaminants were achieved which corresponds to the removal rates of 54%, 85%, 68%, 65%, 68%, 58% and 43% for TDS, turbidity, BOD_5, COD, ammonium-N, nitrate-N and phosphate-P respectively, for the *C. alternifolius* planted VSF CW and 50%, 82%, 70%, 63%, 49%, 54% and 42% for the *C. dactylon* planted system. Hence, *C. alternifolius* and *C. dactylon* planted VSF CWs were shown capable of treating contaminated refinery effluent to discharge permit limits. For most of the parameters considered the performance of the *C. alternifolius* and *C. dactylon* planted VSF CW systems was not significantly different from each other, however, they performed significantly better than the unplanted control.

Keywords: Petroleum refinery wastewater, vertical flow constructed wetlands, contaminants, polishing, planted systems, permit limits.

4.1 Introduction

Refinery effluent refers to wastewater generated from refining crude oil and manufacturing fuels, lubricants and petrochemical intermediates (Diya'uddeen et al., 2011). According to Igunnu and Chen (2012) approximately 40 million m^3 of water are produced daily globally from both oil and gas fields, and more than 40 % of it partially treated is discharged into the environment. The petroleum refinery wastewater produced is characterized by a great range of organic and inorganic pollutants including total dissolved solids (TDS), phenol, oil and grease, ammonia, sulphide and polyaromatic hydrocarbons (Ahmadun et al., 2009; Hoshina et al., 2008). Thus, it is desirable to treat this wastewater to permit limits before discharge into the environment. Some of the treatment methods used are membrane filtration, thermal treatment, biological aerated filters, hydro-cyclones, gas flotation and evaporation ponds (Diya'uddeen et al., 2011). Disadvantages of these treatment technologies are that some of the wastewater generated during backwash and cleaning processes requires further treatment, high level of skilled labour is required, sludge accumulation in the sedimentation basins which can account for up to 40 % of the total cost of the technology, as well as lack of water recovery (Igunnu and Chen, 2012). Simple, low-cost, low energy consumption and effective treatment technologies such as constructed wetlands are desirable, especially for developing countries.

Constructed wetlands (CWs) are integrated eco-systems consisting of shallow ponds or channels planted with aquatic plants, which rely on microbial, biological, physical and chemical processes to purify wastewater containing dissolved or particulate pollutants (Dipu and Thanga, 2009; Imfeld et al., 2009; Mena et al., 2008; Vymazal, 2011). The wetlands are engineered to treat various types of wastewater, ranging from domestic to industrial (Kadlec and Wallace, 2008). In addition, when they are compared with conventional treatment systems, they are ecologically friendly (Jing and Lin, 2004; Ong et al., 2009), have low investment and operation costs (Yang and Hu, 2005) and produce high quality effluent with less dissipation of energy. Asides these, they are relatively simple to operate (Kivaisi, 2001; Song et al., 2006).

CWs can be built in all continents except Antarctica (Vymazal, 2011). They are particularly ideal for tropical and subtropical regions where the climate supports plant growth and microbial activity all year round which enhances the remediation processes (Merkl et al., 2005). Both surface and subsurface CWs are very effective in treating petroleum contaminants to compliance limits, both at small and large scale (Ji et al., 2002; Murray-Gulde et al., 2003; Shpiner et al., 2009; Wallace, 2001). Despite all the benefits of CWs for petroleum wastewater treatment, there is no research available in Nigeria focusing on this treatment system. This study is, therefore, aimed at bridging this gap and to assess for the first time the potentials of subsurface flow CWs for treatment of secondary refinery effluent in Nigeria. Also, conventional treatment methods cannot properly remove all contaminants of concern in the effluents of petroleum refineries. Thus, this study is aimed at using VSF CWs planted with locally available plants (*C. alternifolius* and *C. dactylon*) at the Kaduna refinery discharge point to effectively remove the remaining contaminants in the secondary refinery effluent to meet the compliance limits prior to discharge into the environment as well as to identify indigenous bacteria found in the environment associated with petroleum pollution that can contribute to the treatment of such wastewater under tropical conditions.

4.2   Materials and methods

4.2.1 Description of experimental study site

The study site is located at the Kaduna Refining and Petrochemical Company (Kaduna, Nigeria). The Kaduna state is located in the Northern guinea savannah ecological zone of Nigeria. It lies between Latitude $9\,^0$ N and $12\,^0$ N and Longitude $6\,^0$ E and $9\,^0$ E of the prime meridian. The climatic condition is categorized by constant dry and wet seasons. The rains begin in April/May and ends in October, while the dry season starts late October and stops in March of the subsequent year. The mean annual rainfall is between 1450 - 2000 mm with a mean daily temperature regime ranging from 25 to 43^0 C, and a relative humidity varying between 20 and 40% in January and 60 and 80% in July. It has a solar radiation ranging between 20.0 - 25.0 Wm^{-2} day^{-1} (NIMET, 2010).

The Kaduna Refinery and Petrochemical Company (KRPC) is the third largest refinery in Nigeria (Bako et al., 2008). It has a capacity of 110,000 barrels per stream day (BPSD). The company processes Escravos light crude and Ughelli Quality Control Centre (UQCC) crude oil into fuels and lubes products (Mohammed et al., 2012). In the cause of processing crude oil into finished goods, lots of processes require the use of water for several purposes, hence, wastewater is generated. The company employs a series of physical, chemical and biological treatment methods for the generated wastewater before being released into the discharge channel. The treatment methods include: oil skimming, oxidation, biodegradation, clarification, chemical oxidation, filtration and evaporation. An investigation on the quality of this secondary discharged effluent was conducted from September 2011 to December 2012 to allow the design of the constructed wetlands used in this study.

4.2.2 Refinery effluent sampling

For the purpose of this study, this secondary refinery effluent will be referred to as secondary wastewater. Secondary treated wastewater samples were taken from the VSF CW outlets twice monthly. The containers were thoroughly rinsed three times with the treated wastewater before samples were taken. Samples were collected with into 2 litres labeled polyethylene containers. In-situ measurements were carried out for pH, temperature and turbidity using handheld instruments. A portable HACH conductivity meter was used for electrical conductivity and temperature and a HANNA Instrument LP 2000 turbidity meter was used for turbidity determination. The samples were then placed in an ice-chest and convened to the laboratory for the determination of biological oxygen demand (BOD_5), chemical oxygen demand (COD), total dissolved solids (TDS), total suspended solids (TSS), nitrate-N, ammonium-N, and phosphate-P. These parameters were analyzed according to the procedures described in Standard Methods for the Examination of Water and Wastewater (APHA, 2002): Open reflux, titrimetric method for COD (Maine et al., 2009); 5-Day incubation method for BOD_5 (Maine et al., 2009); gravimetric methods for TDS and TSS, spectrophotometric analysis for phosphate, spectrophotometric analysis for nitrate-N and ammonium-N (HACH, 1997) and total plate counts for bacteriological analysis.

4.2.3 Constructed wetland design and operation

In January 2012, six VSF CWs were developed with cylindrical containers (Fig.4.1), they were set up near the effluent drain channel of KRPC. Each vertical unit had a diameter of 47 cm and a height of 55 cm and a total volume of 95 litres. The media used was gravel mixed with coarse sand. The buckets were filled with a layer of 20 cm of gravel (25 - 36 mm) and coarse sand, followed by 15 cm medium sized gravel (16 - 25 mm) followed by 15 cm of 6 - 10 mm sized gravel particles at the top. Four buckets were planted with *C. alternifolius* and *C. dactylon* found growing freely within the refinery premises. Two buckets without macrophytes served as the control. The schematic diagram of the microcosm study is shown in Fig. 4.2. Fig. 4.2 shows the inflow of the secondary wastewater (influent) and the outflow of the treated wastewater (effluent).

Secondary wastewater was pumped from the refinery drain through a diverting channel using a 5.5 horse power pump into 5000 capacity litre collection tank which subsequently flows into VSF CWs by gravity continuously through perforated polyvinylchloride (PVC) pipe of size of 50 mm installed with control valves that regulate the flow rate (Figs. 4.1 and 4.2) at hydraulic loading rates of 4.65 L/m^2 h and a theoretical hydraulic retention time of 48 hours with a porosity of 0.40 and pore volume of 38L. The control valves were inserted at the outlets of the dosing tank and the VSF CWs. The perforated pipes installed at the inlet and out of the cells were used to enable equal distribution of wastewater in and out of the wetlands. The influent tank (5000 L) was refreshed every 5 to 7 days with secondary wastewater.

4.2.4 Bacteriological sampling and analysis

Influent and effluent samples were collected from the VSF CWs for analysis using standard methods starting in May 2012. Gravel samples were taken at 40 cm depth from the six VSF CWs and plant roots were taken at the same depth from the four planted VSF CWs for total viable bacteria count and identification following the protocol described by Hamza et al. (2012) and Kaur et al. (2010). For the gravel and roots of each plant, the samples were placed in Eppendorf tubes containing distilled water for 1h and then were shaken well. 1 ml from these solutions was mixed in 9 ml sterilized water to make a 10^{-1} serial dilution of this solution, also 1 ml of the wastewater samples collected from each of the wetlands were mixed in 9 ml sterilized water to make a 10^{-1} dilution. 20 g of nutrient agar was put in 1 L graduated flask; it was filled up with sterilized water to the mark. This was autoclaved at 10.342 Kpa and 120 ^0C for 20 min, it was allowed to cool down to 37 ^0C. 1 ml of each of the dilutions was spread evenly on agar-medium Petri dishes to determine the number of populations per gram of samples. Individual colonies of bacteria strains were identified based on culture, morphology and biochemical test using the method described by Hamza et al. (2012) and Kaur et al. (2010).

Fig.4.1. (a) Experimental set up, (b) the macrophytes at start-up, (c) C. dactylon five months after planting and (d) C. alternifolius five months after planting.

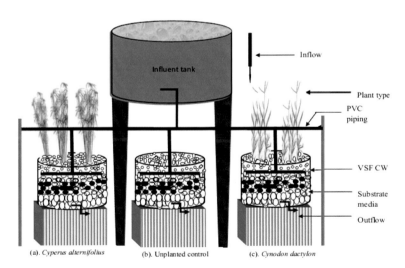

Fig. 4.2. Schematic diagram of experimental set-up showing the inflow (influent) and outflow (effluent) pipes of the different VSF CWs: (a) C. aternifolius, (b) unplanted and (c) C. dactylon

4.2.5 Data analysis

Statistical analysis was performed using the SPSS Statistics 16.0 for Windows (SPSS Inc., Chicago, IL, USA; version 16.0). One way analysis of variance (ANOVA) at 95% significance level was used to examine the performance of the treatment types. Homogeneity of variance tests was done by Levene Statistic and multiple comparisons using Tukey Honest Significant difference (HSD). The Lowest Standard Deviation (LSD) and Duncan Multiple Range (DMR) test were used to compare the means between influent and effluents of the treatment types for the selected parameters of concern.

The treatment efficiency of the system was calculated as the percent of contaminant removal, R for each of the parameters using the following formula:

$$R = \frac{C_i - C_e}{C_i} \times 100 \quad (1)$$

where R is the contaminant removal in percent, C_i is the influent concentration, mg/L and C_e is the effluent concentration, mg/L.

4.3 Results

4.3.1 Treatment efficiencies of the VSF - CWs

Table 4.1 gives the physical, chemical and nutrient qualities of influent (secondary wastewater) and the treated wastewater (effluent) by the VSF CWs planted with *C. alternifolius* and *C. dactylon*. The influent pH had a mean value of 7.3 ± 1.6. The pH of the effluents from the *C. alternifolius* VSF CWs ranged between 6.9 ± 0.3 and 7.0 ± 02 (weakly acidic) and *C. dactylon* VSF CW ranged was averagely 6.9 ± 0.2 (weakly acidic).

The mean influent turbidity concentration (32.1 ± 16.1 NTU) was significantly ($p < 0.05$) higher than the treated effluents (*C. alternifolius*, *C. dactylon* and control VSF CW). *C. alternifolius* planted VSF CWs had values that ranged from 4.8 ± 5.2 and 5.0 ± 5.9 NTU, while effluents from *C. dactylon* planted VSF CWs contained 5.5 ± 7.1 and 6.3 ± 7.5 NTU.

The secondary wastewater had TDS concentrations of 255.5 ± 70.3 mg/L. These concentrations were reduced in all the VSF CWs. In the *C. alternifolius* planted VSF CWs, TDS effluent concentrations were reduced to 116.6 ± 25.8 mg/L, while in the *C. dactylon* VSF CWs, effluent concentration were between 127.0 ± 29.6 mg/L and 128.1 mg/L ± 29.8. The influent TDS concentrations were significantly different from the planted units and the unplanted, but the effluents from the planted units were not significantly different from each other. However, they significantly different from the unplanted.

The mean influent TSS was 60.8 ± 20.2 mg/L. The mean effluent concentrations in *C. alternifolius* and *C. dactylon* planted VSF CWs were between 25.6 ± 6.3 to 26.6 ± 7.9 mg/L and 26.2 ± 5.1 mg/L, respectively. The control unit had TSS effluent concentrations ranging from 36.9 ± 9.5 to 37.3 ± 10.3 mg/L.

The TSS concentrations were thus reduced to below the threshold limit of 30 mg/L; however, this was not achieved in the control units. The mean TSS concentrations were compared statistically, and it showed that the planted VSF CWs were not statistically different from each other, but they significantly different from the unplanted VSF CWs.

BOD_5 concentrations in the influent were 95.3 ± 41.4 mg/L. The effluent concentrations in the *C. alternifolius* planted VSF CWs ranged between 29.1 ± 30.8 to 32.8 ± 35.0 mg/L and *C. dactylon* planted units between 28.7 ± 31.4 to 28.9 ± 25.0 mg/L. The unplanted VSF CWs had BOD_5 concentrations ranging from 48.8 ± 22.8 to 52.0 ± 22.8 mg/L. The Duncan multiple range test was used to compare the BOD_5 of all the treatment systems. There was no significant difference ($p > 0.05$) in the BOD_5 of the *C. alternifolius* and *C. dactylon* planted VSF CWs; however, the planted VSF CWs showed statistically significant difference ($p < 0.05$) in their BOD_5 from that of the unplanted VSF CWs.

The influent mean COD concentration was 164.0 ± 61.9 mg/L. The effluents of the *C. alternifolius* and the *C. dactylon* planted wetlands varied significantly between 55.3 ± 66.8 mg/L and 58.5 ± 65.3 to 62.1 ± 61.1 mg/L, respectively, with lower mean COD concentrations than the unplanted with mean COD concentrations from 95.4 ± 54.3 to 98.7 ± 55.2 mg/L. However, no significant difference was found between the effluent mean COD concentrations of the *C. alternifolius* and *C. dactylon* planted CW systems.

The mean influent nitrate-N and ammonium-N concentrations were 1.6 ± 1.0 mg/L and 1.8 ± 1.6 mg/L, respectively. The mean nitrate-N reduction in *C. alternifolius* and *C. dactylon* planted VSF CWs was not significantly different, although, they showed significant differences from the unplanted VSF CWs. Meanwhile, the mean ammonium-N concentration reduction in *C. alternifolius* planted VSF CW was significantly higher ($p < 0.05$) than *C. dactylon* planted VSF CW and the unplanted VSF CWs. Moreover, also the removal in the *C. dactylon* planted VSF CW was significantly ($p < 0.05$) higher than the unplanted VSF CW.

The mean influent phosphate-P concentrations were 4.0 ± 2.0 mg/L. *C. alternifolius* planted VSF CWs (2.2 ± 1.5 to 2.4 ± 1.5 mg/L) and *C. dactylon* (2.2 ± 1.5 to 2.5 ± 1.8 mg/L) planted VSF CWs showed no significance difference at p < 0.05. The mean phosphate-P concentrations were within the discharge limit throughout the period of investigation.

In this present study, reduction of the TDS concentrations was observed both in the *C. alternifolius* and *C. dactylon* planted VSF CWs with a concentration reduction of 54% and 50%, respectively, while the unplanted cells had a TDS reduction of only 41% as shown in Fig. 4.1. Only effluents from the *C. alternifolius* planted VSF CWs met the turbidity 5.0 NTU discharge permit limit. Yet, all the units including the unplanted VSF CWs had a turbidity removal of over 60% (Fig. 4-3). The primary mechanism for the removal of solids in CWs is sedimentation and filtration.

The average BOD_5 removal efficiency amounted to 68%; 70% and 47% for the *C. alternifolius, C. dactylon* and unplanted VSF CWs, respectively. However, these reductions were not within the threshold limits of 10 mg/L of WHO and FEPA. Meanwhile, average COD removal efficiencies of 65% and 63% were observed for *C. alternifolius* and *C. dactylon* planted VSF CWs, respectively. The unplanted VSF CWs had an average COD removal of 40%.

4.3.2 Properties of bacterial isolates in wastewater, gravel and root media

Bacterial counts showed various genera of aerobic bacteria present in effluent from VSF CWs, gravel and root samples collected from the VSF CWs (Figure 4.2a and Figure 4.2b). The pure cultures of bacteria in the media consisted of rod-shaped and coccoid bacteria (Table 4.2). The cultures were mainly gram-positive (with the exception of two cultures), some cultures were pigmented (purple, yellow, pink and golden yellow) and others were not. The cultures were subjected to four carbohydrate tests. The tests were mainly negative. The biochemical characteristics of the isolates are shown in Table 4.2. The isolates from wetland VSF CWs belong to the families of *Bacillaceae,* comprising 3 species (*B. cereus, B. lichenformis* and *B. subtilis*), *Enterobacteriaceae,* comprising 2 genera *Klebsiella* (i.e *Klebsiella pneumoniae*) and *Escherichia* (*E. coli*) as well as *Micrococcaceae* comprising 2 genera (*Micrococcus* and *Staphylococcus*). These isolates were identified on the basis of their morphology and biochemical properties.

The total number of bacteria in the composite wastewater, gravel and root samples from the six VSF CWs ranged from 3.2 x 10^2 to 2.3 x 10^3 CFU/g wet weight; 5.9 x 10^2 to 7.9 x 10^2 CFU/g wet weight and 4.9 x 10^3 to 7.4 x 10^3 CFU/g wet weight in March 2012, for respectively, wastewater, gravel and root samples. In August 2012, the total number of bacteria found in the collected samples had increased, ranging from 2.2 x 10^4 to 2.2 x 10^5 CFU/g wet weight; 8.0 x 10^4 to 4.8 x 10^5 CFU/g wet weight; 5.7 x 10^5 to 6.8 x10^5 CFU/g wet weight for wastewater, gravel and root samples, respectively. *C. dactylon* (L) planted VSF CWs had the highest bacterial number in its effluent and *C. alternifolius* planted VSF CWs having the least in March, 2012. However, in August, 2012, the unplanted control VSF CW had the lowest bacterial count. Generally, most of the bacteria were found on the roots of the plants in the wetland cells (Fig. 4.2).

Post hoc multiple comparisons were conducted for the total bacteria count found in the wastewater samples and on the gravel and roots of the plants using the Tukey HSD, LSD and Bonferroni test. The total bacteria count in the wastewater in *C. alternifolius* planted VSF CW was not significantly different from *C. dactylon* planted VSF CW and the control VSF CW over time. The results were also tested using a two-way multivariate analysis of variance (MANOVA), which revealed a similar trend where the total bacteria count in the wastewater were not significantly different from each other, irrespective of the time of sample collection (March 2012 and August 2012) and the types of plants used (*C. alternifolius* and *C. dactylon*) as well as the control system (no plant).

The total bacterial count found on the gravel in *C. alternifolius* and *C. dactylon* planted VSF CWs were not statistically different from each other, however, they were significantly different from those found on the gravels in the control cell. Subjecting the results to the Duncan test, the total bacterial count on the gravel of all the VSF CWs varied significantly from each other.

The total bacterial count found on the root samples of *C. alternifolius* and *C. dactylon* plants were not significantly different from each other. At the start-up of the treatment system (March 2012), the *C. alternifolius* planted system had a higher number of bacteria. However, in August 2012; there were more bacteria on the roots of *C. dactylon* plants.

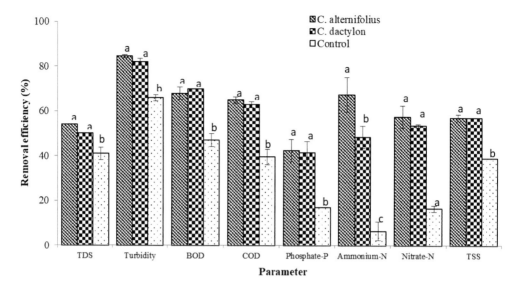

Fig. 4.3. Removal efficiency of wastewater parameters based on initial and final concentrations. Note: Values on the same column with different superscript are significantly different (P ≤ 0.05) while those with the same superscript are not significantly different (P ≥ 0.05) as assessed by LSD, Tukey (HSD) and Duncan's Multiple Range Test.

Table 4.1: Physical and chemical characteristics of influent qualities and effluent of planted and control VSF CWs treating Kaduna refinery wastewater

Treatment types	pH	Temp, °C	Turbidity (NTU)	TDS (mg/L)	TSS (mg/L)	BOD$_5$ (mg/L)	COD (mg/L)	Nitrate-N (mg/L)	Ammonium-N (mg/L)	Phosphate-P (mg/L)
Influent	7.3±1.6[a]	28.4±2.3	32.1±16.1	255.5±70.3	60.8±20.2	95.3±41.4	164.00±61.9	1.62[b] ±1.0	1.81±1.6	4.0[b]±2.0
C.alternifolius1	6.9±0.3	28.0±1.9	5.0[a] ±5.9	116.6±26.9	25.6±6.3	32.8±35.0	59.16[ab]±76.3	0.63[a]±0.7	0.49[a]±0.6	2.4±1.5
C.alternifolius2	7.0±0.2	28.0±1.9	4.8[a] ±5.2	116.6±25.8	26.6±7.9	29.1±30.8	55.28[a] ±66.8	0.74[a]±0.8	0.70[a]±0.8	2.2[a] ±1.5
C.dactylon1	6.9±0.2	28.1±1.8	5.5[ab]±7.1	127.0±29.6	26.2±5.1	28.7±31.4	58.45[ab]±65.3	0.75[a]±0.7	1.00[ab]±1.4	2.2±1.5
C.dactylon2	6.9±0.2	27.9±1.9	6.3[ab]± 7.5	128.1±29.8	26.2±4.7	28.9±25.0	62.09[ab]±61.1	0.77[a]±0.7	0.87[ab]±1.2	2.5±1.8
Control1	7.44±1.1	28.3±2.0	10.6[ab]±9.6	156.8±34.5	36.9[b] ±9.5	48.8[ab]±22.8	95.37[ab]±54.3	1.35[b]±0.9	1.58[bc]±1.3	3.3[ab]±2.0
Control2	7.4±1.0	28.3±2.0	11.3[b]±10.4	161.3[b]±38.7	37.3[b]±10.3	52.0[b]±22.8	98.72[b]±55.2	1.39[b]±1.0	1.90±1.8	3.3[ab]±2.0
WHO&FEPA*	6.0-9.0	40.0	5.0	2000	30	10	40	-	0.2	5.0

a Mean ± standard deviation (SD). Values are Means of two replicate (n=2), Values are Mean ±Standard Deviation. Values on the same column with different superscript are significantly different P≤ 0.05) while those with the same superscript are not significantly different (P ≥ 0.05) as assessed by LSD, Tukey (HSD) and Duncan's Multiple Range test. * WHO (World Health Organization) and FEPA (Federal Environmental Protection Agency, Nigeria) Permissible limits for wastewater discharge

Table 4.2. Biochemical properties of bacterial isolates from the VSFCWs effluent, gravel and roots

Organisms	Pigment	Gram stain reaction	CAT	COUG	SH	CIT	URE	HAE	MR	VP	IND	H$_2$S	CHO L	CHO G	CHO S	CHO F
Bacillus cereus	-	Positive Rod	+	-	+	-	-	-	-	-	-	-	-	-	-	-
Bacillus lichenformis	-	Positive Rod	+	-	+	+	-	-	-	+	-	-	-	-	-	-
Bacillus subtilis	-	Positive Rod	+	-	+	+	-	β	-	+	-	-	-	-	+	-
Escherichia coli	-	Negative Rod	+	-	-	+	-	-	+	-	+	-	+	+	+	-
Klebsiella pneumoniae	-	Negative Rod	+	-	-	+	-	-	-	+	-	+	+	+	-	-
Micrococcus agilis	Purple	Positive Cocci	+	-	-	-	-	-	-	-	-	-	-	-	-	-
Micrococcus luteus	Yellow	Positive Cocci	+	-	-	-	-	-	-	-	-	-	-	-	-	-
Micrococcus roseus	Pink	Positive Cocci	+	-	-	-	-	-	-	-	-	-	-	-	-	+
Staphylococcus roseus	Golden Yellow	Positive Cocci	+	+	-	-	-	-	-	-	-	-	-	+	+	+
Staphylococcus aureus	-	Positive Cocci	+	-	-	-	-	-	-	-	-	-	-	-	-	+
Staphylococcus feacaus	-	Positive cocci	-	-	-	-	-	γ	-	-	-	+	-	+	-	-

Key:
CAT = Catalase test   CIT = Citrate utilization test   MR = Methyl red test   H₂S = hydrogen sulphide
COUG = Coagulase test   URE = Urease Test   VP = Voges Proskauer   G = Glucose Sugar Fermentation test
SH = Starch hydrolysis test   HAE - Haemolysis test   IND = Indole test   S = Sucrose sugar fermentation test
                                                  production test   F = Fructose Sugar Fermentation test

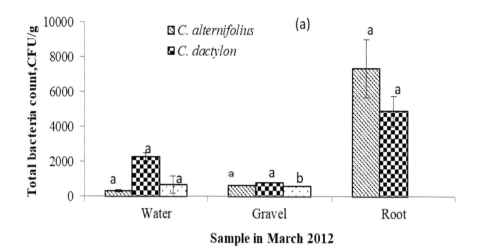

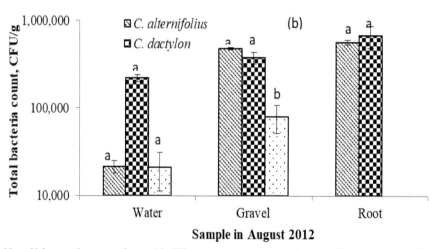

Note: Values on the same column with different superscript are significantly different P≤ 0.05) while those with the same superscript are not significantly different (P ≥ 0.05) as assessed by LSD, Tukey (HSD) and Duncan's Multiple Range Test.

Fig. 4.4. Mean total bacteria count in C. altenifolius and C. dactylon planted and control VSF CWs in

(a) March 2012 and (b) August 2012.

4.4 Discussion

4.4.1 Vertical subsurface flow CW for petroleum secondary effluent treatment

VSF CWs planted with *C. alternifolius* and *C. dactylon* effectively removed the contaminants from the secondary refinery wastewater. Many processes are involved in the removal of these contaminants. For example, the mechanism of suspended matter removal is primarily through physical processes like interception and settling (Mustafa, 2013). Also, BOD$_5$ and COD reduction in wastewater treatment imply organic matter mineralization and metabolism within the treatment system (Hadad et al., 2006; Maine et al., 2009). Generally, there was significant removal of the contaminants of concern in the polished refinery wastewater from the *C. alternifolius* and *C. dactylon* CWs. Effluents from *C. alternifolius* planted VSF CWs varied significantly from *C. dactylon* planted VSF CWs (Table 4.2). However, the unplanted VSF CWs did not vary significantly from the *C. dactylon* planted VSF CWs (Table 4.2).

The BOD$_5$ and COD removal efficiencies achieved in this study is consistent with the removal efficiencies from other studies (Karimi et al., 2014; Kurniadie, 2011). But were higher than the results reported in Kantawanichkul and Wannasri (2013). While Ebrahimi et al. (2013) in their study of the efficiency of *C. alternifolius* achieved a 72% COD reduction in municipal wastewater. In this present study, nitrate-N removal efficiencies of 58%, 54% and 17% for *C. alternifolius*, *C. dactylon* and control cell respectively, were achieved (Figure 4.3). Dissolved oxygen is a limiting factor in the refinery effluent, thus, nitrogen removal pathways from such wastewater in wetlands may be mainly attributed to denitrification, as suggested in the literature (Maine et al., 2006; Garcìa et al., 2010). There was a reduction percent of 43%, 42% and 17% phosphate-P, respectively by the *C. alternifolius*, *C. dactylon* and control treatment VSF CWs. Mustafa (2013) achieved a phosphate reduction efficiency of 52%, indicating that phosphorus removal processes in wetlands were through sorption, precipitation and biomass storage and that the lower phosphate removal efficiencies were due to a lower binding capacity of wetland soil.

From our study, it is evident that planted systems performed better than the unplanted control and this was true for other previous studies as well (Chang et al., 2010; Chen et al., 2006; Jing and Lin, 2004; Maine et al., 2009; Taylor et al., 2011). These observed reduction efficiencies were probably achieved through movement of effluent along the wetlands. In general, the present study showed the good potential of vertical subsurface flow constructed wetlands for the treatment of refinery wastewater.

4.4.2 Bacteriological analysis of subsurface flow constructed wetlands

Microbial distribution and community composition in wetland environments, as well as their physiological and metabolic adaptations are necessary factors in the remediation of oil polluted refinery effluents (Oliveira et al., 2012). The result revealed an increase in the total plate count from March 2012 to August 2012. This may be due to growth of the bacterial population favoured by the conducive wetland environment. In this study, *Micrococcaceae* were found to be the most predominant species, followed by *Bacillaceae, Klebsiella* (*Klebsiella pneumoniae*) and then *Escherichia* (*E. coli*). The presence of these bacteria in the VSF CWs investigated was expected, accordingly, Hamza et al. (2012) recovered *Micrococcaceae* and *Bacillaceae* from Kaduna refinery effluent. Their finding is in agreement with our study in which *Micrococcaceae* were the predominant species in Kaduna refinery effluents. In their studies (Hamza et al., 2012), *Bacillus subtilis* and *Micrococcus luteus* reduced COD by 56.2 % and 52.4 %, respectively. Likewise, the significant different in total bacterial counts in the planted VSF CW systems of this present study may be attributed to biological removal processes (Shabir et al., 2013). Bako et al. (2008) conducted a study to evaluate the potential of *Pseudomonas aeruginosa* and *Penicillium janthinellum* and their mutants to degrade of crude oil in the river Kaduna. The consortia were observed to have a significant decrease in ammonia, nitrate, phosphate and sulphate concentrations. According to Oliveira et al. (2012), these isolated bacteria are found in environments associated with petroleum pollution, suggesting their potential for the remediation of such wastewaters.

4.5 Conclusions

The quality of the refinery effluent treated by VSF CWs was investigated. This study revealed that pH and TDS were within the recommended effluent discharge limits, however, the secondary refinery wastewater had high levels of TSS, BOD_5, COD, turbidity, ammonium-N and phosphate-P. These levels were higher than the permissible levels of discharge allowed by WHO and FEPA. Effluent concentrations were reduced in the *C. alternifolius* and *C. dactylon* planted wetland VSF CWs. Comparing the treatment performances of the planted VSF CWs, *C. alternifolius* and *C. dactylon* planted VSF CWs were significantly not different from each other. However, *C. alternifolius* planted VSF CWs performed better than the *C. dactylon* planted VSF CWs in terms of their reduction of ammonium-N concentrations. Thus, both *C. alternifolius* and *C. dactylon* planted systems are able to reduce the concentrations of contaminants in the Kaduna refinery effluent to the compliance limits set by WHO and FEPA.

4.6 Acknowledgement

The authors acknowledge the management of Kaduna Refinery and Petrochemical Company, Kaduna, Nigeria for giving us the opportunity to conduct this research in their company. The Government of the Netherlands for their financial assistance (NFP-PhD CF7447/2011).

4.7 References

Ahmadun, F.-R., Pendashteh, A., Abdullah, L. C., Biaka, A. D., Madaeni, S. S., Abidin, Z., 2009. Review of technologies for oil and gas produced water treatment. Hazard Mater. 170, 530-551. doi:10.1016/j.jhazmat.2009.05.044.

Akan, J. C., Abdurrahman, F. I., Dimari, G. A., Ogugbuaja, V. O., 2008. Physicochemical determination of pollutants in wastewater and vegetable samples along the Jakara wastewater channel in Kano metropolis, Kano State, Nigeria. Euro J. Sci. Res., 23(1), 122-133. doi:ISSN 1450-216X.

APHA-AWWA-WEF., 2002. Standard Methods for the Examination of Water and Wastewater (20th ed.). Baltimore, Maryland, Maryland, USA: APHA.

Bako, S. P., Chukwunonso, D., Adamu, A. K.., 2008. Bioremediation of refinery effluents by strains of Pseudomonas aerugenosa and Penicillium Janthinellum. Appl. Ecol. Environ. Res., 6(3), 49-60. doi:ISSN 1589 1623.

Chang, N.-B., Xuan, Z., Daranpob, A., Wanielista, M., 2010. A Subsurface Up flow Wetland System for Removal of Nutrients and Pathogens in On-Site Sewage Treatment and Disposal Systems. Env. Eng. Sci., 1-14. doi:10.1089/ees.2010.0087.

Chen, T. Y., Kao, C. M., Yeh, T. Y., Chien, H. Y., Chao, A. C., 2006. Application of a constructed wetland for industrial wastewater treatment: A pilot-scale study. Chemosphere, 64, 497-502. doi:10.1016/j.chemosphere.2005.11.069.

Dipu, S., Thanga, S. G., 2009. Bioremediation of industrial effluents using constructed. 1-6. Retrieved from http://www.eco-web.com/edi/090227.html.

Diya'uddeen, B. H., Wad Daud, M. A., AbdulAziz, A. R., 2011. Treatment technologies for petroluem refinery effluent: A review. Process Saf. Environ Prot., 89, 95-105. doi:10.1016/j.psep.2010.11.003.

Ebrahimi, A., Taheri, E., Ebrampoush, M. H., Nasiri, S., Jalali, F., Soltani, R., Fatehizadeh, A., 2013. Efficiency of constructed wetland vegetated with Cyperus alternifolius applied for municipal wastewater treatment. J. Environ. pub. Health, 1-5.

Garcìa, J., Rousseau, D. P., Morat′o, J., Lesage, E., Matamoros, V., Bayona, J. M., 2010. Contaminant removal processes in subsurface flow constructed wetlands: A Review. Critical Reviews in Environ. Sci. Technol., 40, 561-661. doi:10.1080/10643380802471076.

Gulshan, A. B., Dasti, A. A., 2012 .Physico - chemical nature of oil refinery effluents and it's effects on seed germination of certain plant species. ARPN J. Agric. Biol. Sci., 7(5), 342-345. doi:ISSN 1990-6145.

HACH., 1997. Water Analysis Handbook (3rd ed.). Loveland, Colorado, USA: HACH Company.

Hadad, H. R., Maine, M. A., Bonetto, C. A., 2006. Macrophyte growth in a pilot-scale constructed wetland for industrial wastewater treatment. Chemosphere, 63, 1744-1753. doi:10.1016/j.chemosphere.2005.09.014.

Hamza, U. D., Mohammed, I. A., Sale, A., 2012. Potentials of bacterial isolates in bioremediation of petroleum refinery wastewater. Appl. phytotech. Environ. sanitat., 1(3), 131-138. doi:ISSN 2088-6586.

Hoshina, M. M., de Angelis, D. d., Marin-Morales, M. A., 2008. Induction of micronucleus and nuclear alterations in fish (Oreochromis niloticus) by a petroleum refinery effluent. Mutat. Res., 656, 44-48. doi:10.1016/j.mrgentox.2008.07.004..

Igbinosa, E. O., Okoh, A. I., 2009. Impact of discharge wastewater effluents on physico-chemical qualities of a receiving watershed in a typical rural community. Int. J. Environ.Sci. Tech., 6(2), 175-182.

Igunnu, E. T., Chen, G. Z., 2012. Produced water treatment technologies. Int. J. Low-Carbon Technol., 1-8. doi:10.1093/ijlct/cts049.

Imfeld, G., Braeckevelt, M., Kuschk, P., Richow, H. H., 2009. Review. Monitoring and assessing processes of organic chemicals removal in constructed wetlands. Chemosphere, 74, 349-362. doi:10.1016/j.chemosphere.2008.09.062.

Israel, A. U., Obot, I. B., Umoren, S. A., Mkpenie, V., Ebong, G. A., 2008. Effluents and Solid Waste Analysis in a Petrochemical Company - A Case Study of Petrochemical Company Ltd, Port Harcourt, Nigeria. E-J Chemistry, 5(1), 74-80..

Ji, G. D., Sun, T. H., Qixing, Z., Xin, S., Shijun, C., Peijun, L., 2002. Subsurface flow wetland for treating heavy oil-produced water of the Liaohe Oilfield in China. Ecol. Eng., 18, 459-465. doi:PII: S0925-8574(01)00106-9.

Jing, S., Lin, Y., 2004. Seasonal effect on ammonia nitrogen removal by constructed wetlands treating polluted river water in southern Taiwan. Environ Pol., 127, 291-301. doi:10.1016/S0269-7491(03)00267-7.

Kadlec, R. H., Wallace, S. D., 2008. Treatment Wetlands (2nd ed.). CRC Press. doi:13:978-1-56670-526-4.

Kantawanichkul, S. A., 2013. Wastewater treatment performances of horizontal and vertical subsurface flow constructed wetland systems in tropical climate. Songklanakarin J. Sci. Technol., 35(5), 599-603. Retrieved July 3, 2015, from http://www.sjst.psu.ac.th.

Karimi, B., Ehrampoush, M. H., Jabary, H., 2014. Indicator pathogens, organic matter and LAS detergent removal from wastewater by constructed wetlands. J. Environ Health Sci Eng., 12. doi:10.1186/2052-336X-12-52.

Kaur, A., Vats, S., Rekhi, S., Bhardwaj, A., Goel, J., Tanwar, R. S., Gaur, K. K., 2010. Physico-chemical analysis of the industrial effluents and their impact on the soil microflora. Procedia Environ. Sci., 2, 595-599. doi:10.1016/j.proenv.2010.10.065.

Kivaisi, A. K., 2001. The potential for constructed wetlands for wastewater treatment and reuse in developing countries; a review. Ecol. Eng., 16, 545-560. doi:PII: S0925-854(00)00113-0.

Kurniadie, D., 2011. Wastewater treatment using vertical subsurface flow constructed wetland in Indonesia. American J. Env.Sci, 7(1), 15-19. Retrieved July 2, 2015.

Lawson, E. O., 2011. Physico-chemical parameters and heavy metal contents of water from the mangrove swamps of Lagos Lagoon, Lagos, Nigeria. Adv Biolog. Res., 5(1), 8-21..

Mahre, M. Y., Akpan, J. C., Moses, E. A., Ogugbuaja, V. O., 2007. Pollution indicators in River Kaduna, Nigeria. Trends Appl. Sci. Res., 2(4), 304-311. doi:ISSN 1819-3579.

Maine, M. A., Sune, N., Hadad, H. R., Sanchez, G., Bonetto, C. , 2006. Nutrient and metal removal in a constructed wetland for wastewater treatment from a metallurgic industry. Ecol. Eng., 26, 341-347. doi:10.1016/j.ecoleng.2005.12.004.

Maine, M. A., Sune, N., Hadad, H. R., Sanchez, G., Bonetto, C., 2009. Influence of vegetation on the removal of heavy metals and nutrients in a constructed wetland. Environ. Manage., 90, 355-363. doi:10.1016/j.jenvman.2007.10.004.

Mena, J., Rodriguez, L., Nunez, J., Fernandez, F. J., Villasenor, J., 2008. Design of horizontal and vertical subsurface flow constructed wetlands treating industrial wastewater. WIT Transact. Ecol. Environ., 111(9), 55-557.

Merkl, N., Schultze-Kraft, R., Infante, C. , 2005. Assessment of tropical grasses and legumes for Phytoremediation of petroleum-contaminated soils. Water Air Soil Pol., 165, 195-209.

Mohammed, J., A. D., Auta, M., 2012. Simulation of Kaduna Refining and Petrochemical Company (KRPC) Crude Distillation Unit (CDU I) using Hysys. Int. J. Adv Sci. Res Technol., 1(2), 1-6. doi:ISSN: 2249-9954.

Murray-Gulde, C., Heatley, J. E., Karanfil, T., Rodgers Jr, J. H., Myers, J., 2003. Performance of a hybrid reverse osmosis-constructed wetland treatment system forbrackish oil field produced water. Water Res., 7705-7713. doi:PII: S0043-135(02)00353-6.

Mustafa, A., 2013. Constructed wetland for wastewater treatment and reuse: A case study of developing country. Internat. J. Environ.Sci. Dev, 4(1), 20-24. doi:10.7763/IJESD.2013.V4.296.

NIMET., 2010. Nigeria Meteorological Agency, Nigeria. Kaduna, Kaduna, Nigeria.

Nwanyanwu, C. E., Abu, G. O., 2010. In vitro effects of petroleum refinery wastewater on dehydrogenase activity in marine bacterial strains. Ambi-Agua, Taubatè, 5(2), 21-29. doi:10.4136/ambi-agua.133.

Oliveira, P. F., Vasconcellos, S. P., Angolini, C. F., da Cruz, G. F., Marsaioli, A. J., Santos Neto, E. V., Oliveira, V. M., 2012. Taxonomic Diversity and Biodegradation Potential of Bacteria Isolated from Oil Reservoirs of an Offshore Southern Brazilian Basin. J Pet Environ Biotechnol., 3(7). doi:10.4172/2157-7463.1000132.

Ong, S., Uchiyama, K., D., I., Yamagiwa, K.., 2009. Simultaneous removal of color, organic compounds and nutrients in azo dye-containing wastewater using up-flow constructed wetland. Hazard Mat., 165, 696-703. doi:10.1016/j.jhazmat.2008.10.071.

Shabir, G., Afzal, M., Tahseen, R., Iqbal, S., Khan, Q. M., Khalid, Z. M., 2013. Treatment of oil refinery wastewater using pilot scale fed batch reactor followed by coagulation and sand filtration. American J.Environ. Protect., 1(1), 10-13. doi:10.1269/env-1-2.

Shpiner, R., Liu, G., Stuckey, D. C., 2009. Treatment of oilfield produced water by waste stabilization ponds: biodegradation of petroleum-derived materials. Bioresour Technol., 100, 6229-6235. doi:10.1016/j.biortech.2009.07.005.

Song, Z., Zheng, Z., Li, J., Sun, X., Han, X., Wang, W., Xu, M'., 2006. Seasonal and annual performance of a full-scale constructed wetlands system for sewage treatment in China. Ecol. Eng., 26, 272-282. doi:10.1016/j.ecoleng.2005.10.008.

Taylor, R. C., Hook, B. P., Stein, R. O., Zabinski, A. C., 2011. Seasonal effects of 19 plant species on COD removal in subsurface treatment wetland microcosms. Ecol. Eng., 37(5), 703-710. doi:10.1016/j.ecoleng.2010.05.007.

Vymazal, J., 2011. Constructed wetlands for wastewater treatment: Five decades of experience. Environ. Sci. Technol., 45(1), 61-69. doi:10.1021/es101403q.

Wallace, S. D., 2001. On-site remediation of petroleum contact wastes using subsurface flow wetlands. International Conference on Wetlands and Remediation. Burlington, Vermont..

Yang, L., Hu, C. C., 2005. Treatments of oil-refinery and steel-mill wastewaters by mesocosm constructed wetland systems. Water Sci. Technol., 51(9), 157-164.

Yusuff, R. O., Sonibare, J. A., 2004. Characterization of textile industries' effluents in Kaduna, Nigeria and pollution implications. Global Nest: the Int. J., 212-221.

Ch. 5. Optimization of petroleum refinery wastewater treatment by vertical flow constructed wetlands under tropical conditions: plant species selection and polishing by a horizontal flow constructed wetlands

This chapter was presented and published as:

Hassana Ibrahim Mustapha., 2016. Invasion, management and alternative uses of *Typha latifolia*: A brief review. In: Proceedings of the 2016 Nigeria Institute of Agricultural Engineering Conference, Minna (October 2016).

Hassana Ibrahim Mustapha., Bruggen van J.J.A., P. N. L. Lens., 2018. Optimization of petroleum refinery wastewater treatment by vertical flow constructed wetlands under tropical conditions: plant species selection and polishing by a horizontal flow constructed wetland. Journal of Water, Air & Soil Pollution. 229(4):137-154. DOI: 10.1007/s11270-018-3776-3

Abstract

Typha latifolia-planted vertical subsurface flow constructed wetlands (VSSF CWs) can be used to treat petroleum refinery wastewater. This study evaluated if the removal efficiency of VSSF CWs can be improved by changing the plant species or coupling horizontal subsurface flow constructed wetlands (HSSF CWs) to the VSSF CW systems. The VSSF CWs had a removal efficiency of 76% for biological oxygen demand (BOD_5), 73% for chemical oxygen demand (COD), 70% for ammonium-N (NH_4^+-N), 68% for nitrate-N (NO_3^--N), 49% for phosphate (PO_4^{3-}-P), 68% for total suspended solids (TSS) and 89% for turbidity. The HSSF CWs planted with *T. latifolia* further reduced the contaminant load of the VSSF CW-treated effluent, giving an additional removal efficiency of 74, 65, 43, 65, 58, 50 and 75% for, respectively, BOD_5, COD, NH_4^+-N, NO_3^--N, PO_4^{3-}-P, TSS and turbidity. The combined hybrid CW showed, therefore, an improved effluent quality with overall removal efficiencies of, respectively, 94% for BOD_5, 88% for COD, 84% for NH_4^+-N, 89% for NO_3^--N, 78% for PO_4^{3-}-P, 85% for TSS and 97% for turbidity. *T. latifolia* thrived well in the VSSF and HSSF CWs, which may have contributed to the high NH_4^+-N, NO_3^--N and PO_4^{3-}-P removal efficiencies. *T. latifolia*-planted VSSF CWs showed a higher contaminant removal efficiency compared to the unplanted VSSF CW. *T. latifolia* is thus a suitable plant species for treatment of secondary refinery wastewater. Also a *T. latifolia*-planted hybrid CW is a viable alternative for the treatment of secondary refinery wastewater under the prevailing climatic conditions in Nigeria.

Keywords: Optimization, *Typha latifolia*, Refinery wastewater, Tropics, Hybrid CWs, Discharge limits

5.1 Introduction

Industrialization is linked to major pollution sources of hazardous pollutants into water bodies, especially in developing countries where untreated or only partially treated industrial wastewater is discharged into the environment (Sepahi et al., 2008). This poses considerable environmental problems, because most of the people's livelihood in the pollution affected communities depends on the water bodies for fishing, domestic use and irrigation (Agbenin et al., 2009; Senewo, 2015). There is an increasing awareness of the need for the discharge of well treated wastewater into the environment (Adomokai and Sheate, 2004), especially with the recent agitations for the demand for compensations by oil pollution affected communities in Nigeria based on human and environmental rights (Isumonah, 2015; Senewo, 2015). Senewo (2015) reported that their struggle for survival was more an ecological than a political one.

The petroleum refining industry mainly uses conventional methods to treat the wastewater generated by the refining of crude oil to fuel and other finished products. The most prominent problem of the petroleum industry is the disposal of produced water (Hagahmed et al., 2014). These conventional methods are, however, energy intensive, expensive and generate by-products that are often toxic to both humans and the environment (Ojumu et al., 2005). Hence, there is a need for economically and ecologically friendly refinery wastewater treatment technologies. Constructed wetland technologies may offer such lower construction and maintenance costs for wastewater treatment, which are especially suitable for developing countries (Kivaisi, 2001; Kaseva, 2004; Mustapha et al., 2015). Also, many refinery locations are in the tropics, where constructed wetlands (CWs) are well suitable due to the high temperatures which enhance biodegradation activities (Kantawanichkul et al., 1999).

Wetland plant species are capable of removing many contaminants from the wastewater (Ji et al., 2007; Mustapha et al., 2015). In addition, wetland plant roots provide a habitat that is conducive for the growth of a great diversity of microbial communities which enhance the pollutant removal efficiencies in CWs (Abou-Elela and Hellal, 2012). Mustapha et al. (2015) reported that *Cyperus alternifolius* and *Cynodon dactylon* (L.) Pers. planted VSSF CWs removed 68, 65, 68, 58 and 43% and 70, 63, 49, 54 and 42% of BOD_5, COD, NH_4^+-N, NO_3^--N and PO_4^{3-}-P, respectively, from secondary refinery wastewater. Ji et al. (2007) used *Phragmites australis* to remove COD and BOD_5 in heavy oil-produced wastewater in two reed beds. The treatment showed removal efficiencies of 80 and 88% for COD and BOD_5, respectively, for reed bed no. 1 and 71 and 77% for COD and BOD_5, respectively for reed bed no. 2. The reed beds operated at COD loading rates of 13.3 and 26.7 g m^{-2} d^{-1} corresponding to a hydraulic loading rate (HLR) of 18.75 and 37.5 m^3 day^{-1} and hydraulic retention time (HRT) of 15 and 7.5 days, respectively, for reed bed no. 1 and reed bed no. 2. A hybrid CW is a combination of two or more CWs in series that combines the advantages of the various CW systems to provide a better effluent quality than a single CW system (Ávila et al., 2013; Zurita and White, 2014; Zapater-Pereyra et al., 2015). In addition, they combine aerobic and anaerobic properties of VSSF and HSSF CWs to enhance the removal of COD, phosphorus and, especially nitrogen (Vymazal, 2005; Mena et al., 2008). Hybrid CWs are effective in the treatment of contaminants in various types of wastewater. For instance, Murray-Gulde et al. (2003) achieved a suitable irrigation water from treating blackish produced water with a hybrid reverse osmosis-constructed wetland. The pilot-scale reverse osmosis-constructed wetland system effectively decreased conductivity by 95% and total dissolved solids (TDS) by 94% in the produced water. Herrera Melián et al. (2010) used a 2-stage hybrid CW consisting of VSSF CW as the first stage and HSSF CW as the second stage to treat BOD$_5$, COD, ammonia-N (NH$_4^+$-N), suspended solids (SS), and PO$_4^{3-}$-P in urban wastewater in the Canary Islands (Spain). The main objective of their study was to compare the effect of planting, substrate type (gravel and lapilli) and hydraulic loading rate. Both the gravel and lapilli hybrid CWs showed similar average removal efficiencies exceeding 86, 80, 88, 96, 96 and 24%, for BOD$_5$, COD, NH$_4^+$-N, SS, turbidity and PO$_4^{3-}$-P, respectively. Their study demostrated that hybrid CWs are a robust configuration for wastewater treatment. Ávila et al. (2013) demonstrated the effectiveness of a 3-stage hybrid CW consisting of a VSSF CW, a HSSF CW and a freewater surface wetland connected in series for the treatment of TSS, COD, BOD$_5$, NH$_4^+$-N, TN and PO$_4^{3-}$-P. The 3-stage hybrid CW system had mean removal rates of 97% TSS, 78% COD, 91% BOD$_5$, 94% NH$_4^+$-N and 46% TN.

Improving the treatment performance of wetland systems critically depends on the selection of optimal environmental and operating conditions, plant species and effective use of wetland media (Saeed and Guangzhi, 2012). This present study is aimed at optimizing the performance of previous work on *Cyperus alternifolius* and *Cynodon dactylon*-planted VSSF CWs that have already shown to be able to treat the pollutants present in secondary Kaduna refinery wastewater (Mustapha et al., 2015). The objective of this study was to evaluate the removal efficiency of a *Typha latifolia*-planted VSSF CWs. In addition, it was assessed if the treatment efficiency can be further improved choosing *T. latifolia* as the macrophyte of interest and/or by adding a HSSF CW as an effluent post treatment. The removal of BOD$_5$,

COD, NH_4^+-N, NO_3^--N and PO_4^{3-}-P in the secondary treated refinery wastewater was investigated.

5.2 Materials and methods

5.2.1 Description of experimental study site

This study was conducted at the Kaduna Refinery and Petrochemical Company (Kaduna, Nigeria), which lies between latitude 9^0 N and 12^0 N and longitude 6^0 E and 9^0 E within the Northern guinea savannah ecological zone of Nigeria. Kaduna (Nigeria) has a tropical climatic condition with a mean daily temperature between 25 and 43°C and average annual rainfall between 1450 and 2000 mm (NIMET, 2010). The refinery treats its effluents by chemical addition, clarification, oxidation, oil skimming, filtration and evaporation before being discharged via drainages into the Romi River. More details on the treatment process and characteristics of the petroleum refinery effluent are given in Chapter 4.

5.2.2 Experimental design

Refinery wastewater was pumped from the refinery effluent drain through a diverting channel using a 4.1 kW power pump into a 5 m^3 collection tank, which subsequently flowed gradually by gravity into the wetland cells while the treated effluent was collected at the outlet (Fig. 5.1). PVC pipes with 50 mm diameter were used for the plumbing in and between the wetlands. The configuration of the hybrid CW system for this study (Fig. 5.2) was such that the influent (secondary refinery wastewater), that is the VSSF influent, flows into the VSSF CWs, whereas the effluent from the VSSF CWs is referred to as the HSSF CW influent and the effluent from the HSSF CWs is referred to as final effluent (Fig. 5.2). Treated samples from the outlet of the wetland cells were collected every 2 weeks for both field and laboratory analysis. The microcosm-scaled subsurface flow constructed wetland systems were composed of four VSSF - HSSF hybrid wetlands connected in series (Fig.5.1 and Fig. 5.2).

The VSSF wetlands were circular in shape and made of plastic (47 cm diameter, 55 cm height), while the HSSF wetlands were rectangular in shape (110 cm x 70 cm x 40 cm for, respectively, length, width and depth) and also made of plastic. More details on the design of the VSSF CWs are given in Mustapha et al. (2015). The HSSF wetland cells had an effective volume of 123 L with a porosity of 0.40. The HSSF wetland had a designed flow rate of 0.83 L/h, a hydraulic loading rate of 1.08 L/m^2 h and a hydraulic retention time of 148 hours (Table 5.1). The media types used for the HSSF wetlands was gravel with coarse sand.

Coarse size gravel of 25 - 36 mm was used near the inlet and outlet of the HSSF wetland cells and the middle parts were filled with 6 - 10 mm gravel to support the plant roots. The bottom of both the VSSF and HSSF CWs was fitted with perforated PVC pipes (diameter 50 mm) about 10 cm above the media. These PVC pipes were connected to the collection chamber.

 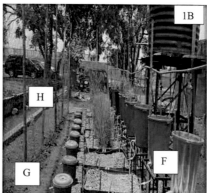

Fig. 5.1a: Startup of experiment in February 2012 (t = 0), b: Experimental setup after 90 days of operation. A – VSSF CW; B – HSSF CW; C - T. latifolia; D - C. alternifolius, E- C. dactylon, F – control VSSF and HSSF CW, G – outlet and H – effluent discharge channel

Two VSSF - HSSF hybrid wetlands were planted with *T. latifolia* 10 cm below the media. Two other unplanted VSSF - HSSF hybrid wetlands served as control to assess the performance of the *T. latifolia*-planted CW. The *T. latifolia* used in this study was collected from a swampy area outside the refinery. *T. latifolia* was used in this study to compare its performance with that of VSSF CW planted with *C. alternifolius* and *C. dactylon* investigated in a previous study (Mustapha et al., 2015).

Table 5.1. Description of microcosm-scale VSSF - HSSF hybrid constructed wetlands used in this study

Property	Description	
Type of constructed wetlands	Vertical subsurface flow (VSSF)	Horizontal subsurface flow (HSSF)
No. of hybrid wetlands	4 (2 planted and 2 unplanted)	4 wetlands (2 planted and 2 unplanted)
Dimension (cm)	47 × 55 (Ø × h)	110 × 70 × 40 (L × W × D)
Substrate	Gravel mixed with coarse sand	Gravel mixed with coarse sand

Loading method	Continuous	Continuous
Design flow rate	0.83 L/h	0.83 L/h
Hydraulic Retention Time	48 h	148 h

5.2.3 Sample collection

Sampling of wastewater started 3 months after the *T. latifolia* transplant into the VSSF and HSSF CWs to allow proper plant establishment, acclimatization and possible biofilm establishment (Kaseva, 2004). Subsequently, influent flowing into the VSSF and HSSF CWs and treated effluent were collected every 2 weeks for both field [temperature, pH, and turbidity] and laboratory [NH_4^+-N, NO_3^--N, PO_4^{3-}-P, BOD_5, COD and TSS] analysis.

5.2.4 Monitoring the growth of *Typha latifolia* in VSSF and HSSF CWs

At the beginning of the experiment, the plant height and number of live shoots of *T. latifolia* were recorded at the time of the transplant and subsequently, every 3 months for a period of 9 months in order to monitor the growth rate of the *T. latifolia* in secondary treated refinery wastewater.

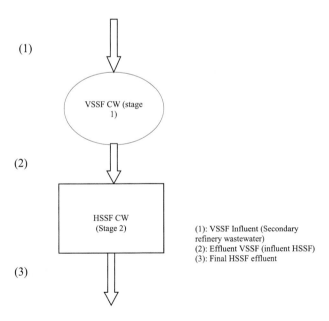

Fig. 5.2. Configuration of hybrid constructed wetlands treating secondary refinery wastewater

5.2.5 Analysis

Temperature, pH and turbidity were measured in the field using a HORIBA pH meter for temperature and pH and a HANNA Instrument LP 2000 for turbidity. TSS measurements were conducted by gravimetric methods dried at 105°C. BOD_5 tests were determined by the 5-day incubation method, COD was determined by the open reflux titrimetric method, NH_4^+-N, NO_3^--N and PO_4^{3-}-P were determined by spectrophotometric methods. All laboratory analyses were carried out using standard methods (APHA, 2002) as described in Chapter 3.

5.2.6 Performance of constructed wetlands based on mass balance calculations

The effects of rainfall and evapotranspiration on pollutant mass load removal were considered. The treatment efficiency of tropical CWs is better evaluated based on mass balance estimations rather than the differences between the inflow and outflow concentrations (Katsenovich et al., 2009; Dhulap et al., 2014). Therefore, the treatment efficiency of these systems was calculated as the percent of load removal efficiency (RE) for each of the parameters considering the following formula from Bialowiec et al. (2014):

Evapotranspiration (ET) was estimated according to Eq. (5.1):

$$ET = Q_{inf} - Q_{eff} \qquad (5.1)$$

where Q_{inf} is the total amount of wastewater entering the wetlands and Q_{eff} is the total amount of wastewater leaving the wetlands (m^3/h). Q_{inf} is computed using Eq. (5.2):

$$Q_{inf} = Q_p + Q_s \qquad (5.2)$$

where Q_p is the amount of precipitation (m^3/h) and Q_s is the amount of water added to the wetland (m^3/h). Therefore, the removal efficiency (RE) of BOD_5, COD, TSS, NH_4^+-N, NO_3^--N and PO_4^{3-}-P was calculated based on loads as given in Eq. (5.3):

$$RE = \frac{(Q_{inf} * C_{inf}) - (Q_{eff} * C_{eff})}{(Q_{inf} * C_{inf})} * 100 \qquad (5.3)$$

where Q_{inf} and Q_{eff} are as previously defined, C_{inf} and C_{eff} are the mean concentrations (mg/L) of a compound in the influent and effluent, respectively. $Q_{inf}*C_{inf} = M_{inf}$ (mg) and $Q_{.eff} * C_{eff} = M_{eff}$ (mg/day) are, respectively, the inflow and outflow mass load.

The contribution made by the HSSF CWs (R_a) is estimated as mean total concentration (C_t) of the wastewater pollutant from the HSSF CWs divided by the total mean pollutant concentration (C_i) in the influent prior to passing through the VSSF CW multiplied by 100%:

$$R_a = \frac{C_t}{C_i} \times 100\% \qquad (5.4)$$

5.3 Results

5.3.1 Growth of *Typha latifolia* in wetlands treating secondary refinery wastewater

T. latifolia growing in the CWs showed a positive growth response in the secondary refinery wastewater throughout the period of experimentation. At the startup of the experiment (day 0), *T. latifolia* was 40 cm in height on average in both the VSSF CW and HSSF CW of the hybrid system (Figs. 5.1 and 5.3). By day 270, it had increased to 113 cm in height in the VSSF CW and 181 cm in the HSSF CW. Similarly, the VSSF CW was started with 11 live shoots of *T. latifolia* and by day 270 they had increased to averagely 136 shoots, while the HSSF CW started with 34 live shoots and increased to 408 live shoots by day 270 (Fig. 5.3). Despite being grown in secondary refinery wastewater for over 270 days, *T. latifolia* did not show any withering or other signs of plant disease.

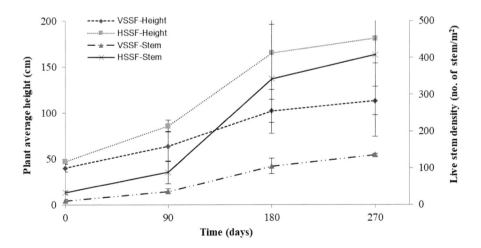

Fig. 5.3. Development of Typha latifolia in VSSF and HSSF hybrid constructed wetland systems fed with secondary refinery wastewater from Kaduna refinery (Nigeria); mean ± SD, n = 2 wetlands.

5.3.2 Physicochemical wastewater parameters

The mean concentrations, standard deviations, minimum and maximum values of the physicochemical parameters of the VSSF influent, *T. latifolia* VSSF effluent and the final *T. latifolia* HSSF effluent during the 308 days period investigated are presented in Table 5.2.

The pH of the VSSF influent ranged from 5.9 to 11.3 (7.4 ± 1.6), the *T. latifolia* VSSF effluent ranged from 6.4 to 7.1 (6.9 ± 1.5) and the final *T. latifolia* HSSF effluent ranged from 6.4 to 7.5 (7.0 ± 1.5). The effluent pH from the control VSSF and HSSF CWs ranged from 6.0 to 9.0 (7.4 ± 1.9) and from 6.1 to 9.9 (7.5 ± 1.9), respectively.

The temperature for the VSSF influent ranged from 24.5 to 32.4 (28.4 ± 2.3) °C, the *T. latifolia* VSSF CWs effluents and the final *T. latifolia* HSSF CWs effluents ranged from 25.0 to 30.2 (27.8 ± 6.1) °C and from 25.20 to 30.1 (27.5 ± 5.9) °C, respectively. The effluents from the control VSSF and HSSF CWs had a temperature ranging from 25.1 to 31.5 (28.3 ± 6.2) °C and from 25.0 to 31.1 (28.1 ± 6.1) °C, respectively.

5.3.3 Turbidity

The VSSF influent had a turbidity of 10.3 to 65.0 NTU. The *T. latifolia* hybrid and control hybrid CWs had a turbidity removal efficiency of 96.9 (± 2.8) and 83.5 (± 11.4)%, respectively, achieving at some stages 100 and 97% turbidity removal for, respectively, *T. latifolia* hybrid and the unplanted control hybrid CWs. The effluent from the *T. latifolia* VSSF and the final effluent from the *T. latifolia* HSSF CWs had a turbidity of, respectively, 4.5 and 1.9 NTU. The control VSSF and HSSF CWs effluent had a turbidity of 10.9 and 7.6 NTU, respectively. The *T. latifolia* planted VSSF and the HSSF CWs thus showed a mean turbidity removal efficiency of 88.6 (±9.5) and 74.6 (± 12.6)%, respectively.

Table 5.2. Composition of secondary refinery wastewater (influent) and effluents from *T. latifolia*-planted VSSF and HSSF and unplanted VSSF and HSSF control constructed wetlands (n = 4)

Parameter	Samples	Mean	Standard deviation	Minimum	Maximum	[a] Discharge standard (FEPA)
Turbidity	VSSF influent	32.1	16.1	10.3	65.0	
(NTU)	*T. latifolia* VSSF effluent	4.5	5.7	0.2	24.0	
	Final *T. latifolia* HSSF	1.9	2.9	0.0	10.3	

Parameter	Treatment					
	effluent Control VSSF	10.9	9.7	1.8	36.9	
	Control HSSF	7.6	7.6	0.8	28.1	5
TSS	VSSF influent *T. latifolia*	60.8	20.2	21.0	99.2	
	VSSF effluent Final *T. latifolia*	19.5	7.3	10.7	43.5	
	HSSF effluent Control	2.1	5.6	4.7	30.5	
	VSSF Control	37.1	9.9	21.6	63.2	
	HSSF	30.0	7.9	19.7	52.1	30
BOD$_5$	VSSF influent *T. latifolia*	95.2	41.4	29.4	190.5	
	VSSF effluent Final *T. latifolia*	24.1	23.5	6.0	88.9	
	HSSF effluent Control	9.3	13.3	0.6	44.6	
	VSSF Control	49.2	21.9	21.8	97.9	
	HSSF	40.1	29.1	15.4	95.9	10
COD	VSSF influent *T. latifolia*	164.0	61.9	60.0	300.1	
	VSSF effluent Final *T. latifolia*	52.6	72.7	11.9	252. 0	
	HSSF effluent Control	30.8	50.7	0.5	182.2	
	VSSF Control	97.0	54.5	44.4	250.8	
	HSSF	63.0	41.2	29.2	160.7	40
Ammonium-N	VSSF influent *T. latifolia*	1.8	1.59	0.3	6.0	
	VSSF effluent	0.9	1.4	0.0	3.9	

	Final *T. latifolia* HSSF effluent	0.5	0.5	0.0	1.7	
	Control VSSF	1.7	1.5	0.2	4.9	<15[b]
	Control HSSF	1.3	1.3	0.1	4.1	
Nitrate-N	VSSF influent *T. latifolia*	1.6	1.0	0.3	4.2	
	VSSF effluent	0.6	0.6	0.0	1.9	
	Final *T. latifolia* HSSF effluent	0.2	0.3	0.0	1.1	< 20[b]
	Control VSSF	1.4	0.9	0.2	3.4	
	Control HSSF	0.7	0.5	0.1	1.7	
Phosphate-P	VSSF influent *T. latifolia*	4.0	2.0	1.1	6.9	
	VSSF effluent	2.1	1.5	0.8	5.8	
	Final *T. latifolia* HSSF effluent	1.3	1.2	0.1	3.9	5
	Control VSSF	3.3	2.0	0.8	6.3	
	Control HSSF	2.6	1.9	0.6	5.7	

FEPA: Federal Environmental Protection Agency (FEPA), Nigeria discharged limit. b (Masi and Martinuzzi, 2007)

5.3.4 Total suspended solids

The VSSF influent had a range of TSS concentrations from 21.0 to 99.2 (60.8 ± 20.2) mg/L. The effluent of the *T. latifolia* VSSF had a concentration range from 10.7 to 43.5 (19.4 ± 7.3) mg/L and the final HSSF CWs effluents had a lower TSS range from 4.7 to 30.5 (12.1 ± 5.6) mg/L. The control VSSF CWs had TSS concentrations from 21.6 to 63.2 (37.1 ± 9.9) mg/L and the control HSSF CWs had a TSS range from 19.7 to 52.1 (30.0 ± 7.9) mg/L (Table 5.2). The TSS concentrations of the effluents from the *T. latifolia* VSSF were lower than the TSS concentrations from the VSSF influent with the final effluents from the *T. latifolia* HSSF

CWs showing a further reduction in TSS concentration, which was below the TSS effluent threshold limit of 30 mg/L for effluent discharge by the WHO and FEPA (Table 5.2).

The VSSF and the HSSF CWs had TSS removal efficiencies of 68.4 (± 8.9) and 50.0 (± 9.0) %, respectively. The *T. latifolia* hybrid and control hybrid CWs had a TSS removal efficiency of 84.5 (± 4.04) and 60.1 (± 9.6)%, respectively, with both systems reaching up to 91 and 74% as maximum TSS removal efficiency.

5.3.5 Biological and chemical oxygen demand

The VSSF influent BOD_5 concentrations ranged from 29.4 to 190.5 (95.3 ± 41.4) mg/L. The effluents from the *T. latifolia* VSSF and the final effluent from the HSSF CWs had BOD_5 concentrations from 6.0 to 88.9 (24.1 ± 23.5) mg/L and 0.6 to 44.6 (9.30 ± 13.3) mg/L, respectively. The control VSSF and HSSF CWs had effluent BOD_5 concentrations from 21.8 to 97.9 (49.2 ± 21.9) mg/L and 15.4 to 95.9 (40.1 ± 29.1) mg/L, respectively.

The VSSF influent COD concentrations ranged from 60.0 to 300.1 (164.0 ± 61.9) mg/L. The *T. latifolia* VSSF effluent and final effluents from the HSSF CWs had COD concentrations from 11.9 to 252.0 (52.6 ± 72.7) mg/L and 0.5 to 182.0 (30.8 ± 50.7) mg/L, respectively. The control VSSF and HSSF CWs had COD concentrations from 44.4 to 250.8 (97.0 ± 54.5) mg/L and 29.2 to 160.7 (63.0 ± 41.3) mg/L, respectively.

The BOD_5/COD ratio of the secondary refinery wastewater was 0.58, this was decreased with time in the *T. latifolia*-planted VSSF CW to 0.46 and further decomposition of organic matter reduced the BOD_5/COD ratio to 0.30 in the *T. latifolia*-planted HSSF CW. Meanwhile, there was a lower biodegradability in the control VSSF CW with a BOD_5/COD ratio of 0.51 and an increased of the BOD_5/COD ratio of 0.64 in the control HSSF CW.

The *T. latifolia* hybrid CWs had an average mass BOD_5 removal efficiency of 93.6 (±7.4)%. From day 238 to day 308, the *T. latifolia* hybrid CWs achieved a 99% BOD_5 removal efficiency (Fig. 5.4a). In contrast, the control hybrid CWs achieved an average BOD_5 removal efficiency of only 65.3 (± 20.5)% (Fig. 5.4b).

The mass influent BOD_5 loading rate varied between 135.5 and 878.2 g m^{-2} d^{-1}, the mass BOD_5 removal rates for the *T. latifolia* VSSF CWs varied between 27.7 and 410.0 g m^{-2} d^{-1} and between 0.6 and 48.1 g m^{-2} d^{-1} for the *T. latifolia* HSSF CWs. For the control VSSF CWs, the mass BOD_5 removal rates varied between 100.4 and 451.1 g m^{-2} d^{-1}, whereas they varied between 16.7 and 103.5 g m^{-2} d^{-1} for the control HSSF CWs.

The mass removal rates for the VSSF influent COD varied between 276.6 and 1383.5 g m^{-2} d^{-1}, and it varied between 55.0 and 1161.7 g m^{-2} d^{-1} for the *T. latifolia* VSSF CWs effluent, whereas the mass removal rates for the effluent from the *T. latifolia* HSSF CWs varied between 0.5 and 196.8 g m^{-2} d^{-1}. The mass removal rates for the control VSSF CWs effluent varied between 204.5 and 1156.1 kg m^{-2} d^{-1} and between 31.5 and 173.5 kg m^{-2} d^{-1} for the control HSSF CWs effluent. Hence, the *T. latifolia* hybrid CWs had an average mass COD removal efficiency of 87.9 (± 13.6)% and occasionally achieved a 99% removal efficiency (Fig. 5.5a). Meanwhile, the control hybrid CWs achieved an average removal efficiency of 71.4 (± 11.0)% (Fig. 5.5b).

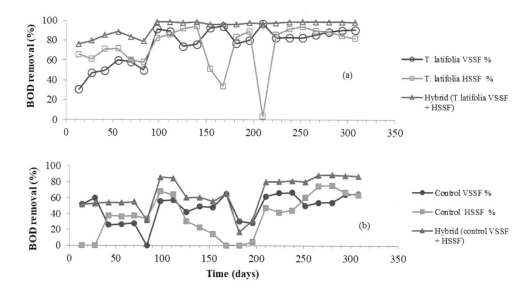

Fig. 5.4. BOD removal efficiency of CW treating Kaduna refinery wastewater: (a) T. latifolia hybrid CW and (b) unplanted control hybrid CW. Mean values and standard deviations are presented in Table 5.2.

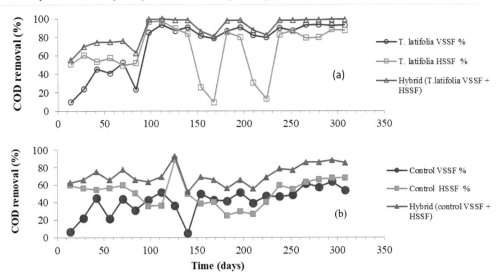

Fig. 5.5. COD removal efficiency of CW treating Kaduna refinery wastewater: (a) T. latifolia hybrid CW and (b) control hybrid CW. Mean values and standard deviations are presented in Table 5.2.

5.3.6 Nitrogenous compounds

The NH_4^+-N concentrations in the VSSF influent ranged from 0.3 to 6.0 (1.8 ± 1.6) mg/L. The VSSF effluent and final HSSF effluent concentrations ranged from, respectively, 0.02 to 3.9 (0.9 ± 1.4) mg/L and 0.0 to 1.7 (0.5 ± 0.5) mg/L. The effluents from the control VSSF and HSSF CWs had NH_4^+-N concentrations ranging from, respectively, 0.2 to 4.9 (1.7 ± 1.5) mg/L and 0.1 to 4.1 (1.3 ± 1.3) mg/L (Table 5.2).

The NO_3^--N concentrations in the VSSF influent ranged from 0.3 to 4.2 (1.6 ± 1.0) mg/L. The VSSF effluent and final HSSF effluent concentrations from the *T. latifolia*-planted VSSF and HSSF CWs ranged from, respectively, 0.02 to 1.9 (0.6 ± 0.6) mg/L and 0.01 to 1.1 (0.2 ± 0.3) mg/L. The effluents from the control VSSF and control HSSF CWs had effluent concentrations ranging from, respectively, 0.2 to 3.4 (1.4 ± 0.9) mg/L and 0.06 to 1.7 (0.7 ± 0.5) mg/L.

High NH_4^+-N removal efficiencies were achieved by the *T. latifolia* VSSF CWs (70 ± 31.7%) and the *T. latifolia* hybrid CWs (83.7 ± 14.8%). Also the nitrate removal efficiency was good from both the VSSF CWs (67.7 ± 23.0 %) and the hybrid system (89.1 ± 12.7%). In some occasions, the *T. latifolia* hybrid CWs achieved a 100% removal efficiency for both NH_4^+-N and NO_3^--N (Fig. 5.6a and 5.7b). For the control hybrid CWs, the overall average removal efficiencies for NH_4^+-N and NO_3^--N were, respectively, 45.3 ($\pm$ 31.8) and 69.3 ($\pm$ 17.5) %.

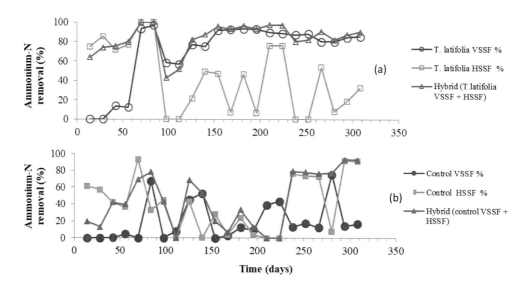

Fig. 5.6. Ammonium-N removal efficiency of CW treating Kaduna refinery wastewater: (a) T. latifolia hybrid CW and (b) control hybrid CW. Mean values and standard deviations are presented in Table 5.2.

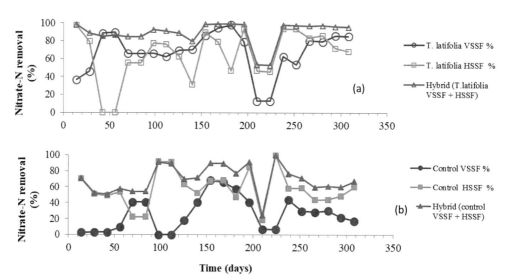

Fig. 5.7. Nitrate-N removal efficiency of CW treating Kaduna refinery wastewater: (a) T. latifolia hybrid CW and (b) control hybrid CW. Mean values and standard deviations are presented in Table 5.2.

5.3.7 Phosphate

The PO_4^{3-}-P concentration in the influent ranged from 1.1 to 6.9 (4.0 ± 2.0) mg/L, whereas the VSSF effluent and the final HSSF effluent from the *T. latifolia* VSSF CWs and HSSF CWs had a PO_4^{3-}-P concentration ranging from 0.8 to 5.8 (2.1 ± 1.5) mg/L and from 0.09 to 3.9 (1.3 ± 1.2) mg/L, respectively. The effluents from the control VSSF and HSSF CWs had PO_4^{3-}-P concentrations ranging from 0.8 to 6.3 (3.3 ± 2.0) mg/L and from 0.6 to 5.7 (2.6 ± 1.9) mg/L, respectively (Table 5.2).

The mean PO_4^{3-}-P removal efficiency of 67.7 (± 23.0%) was observed for *T. latifolia* VSSF CWs, while the *T. latifolia*-planted hybrid CWs had a mean PO_4^{3-}-P removal efficiency of 77.7 (± 12.7)%, whereas the control hybrid CWs had a mean removal efficiency of 53.5 (± 12.5)%. Maximum PO_4^{3-}-P removal efficiencies of 96 (days 210 to 224) and 72% (day 182) were observed for, respectively, the planted and control VSSF CWs (Figs. 5.8a and b).

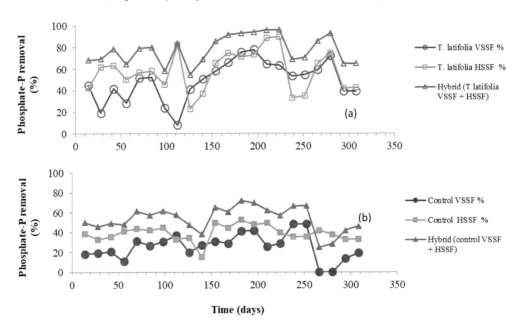

Fig. 5.8. Phosphate-P removal efficiency of CW treating Kaduna refinery wastewater: (a) T. latifolia hybrid CW and (b) control hybrid CW. Mean values and standard deviations are presented in Table 5.2.

5.3.8 Relationship between organic mass loading rates and mass removal rates

Fig. 5.10 and Fig. 5.11 present the relationship between the applied mass loading rates (MLRs) and the corresponding mass removal rates (MRRs) for BOD_5 and COD in *T. latifolia*-planted and control wetland systems. The mean BOD_5 MRRs (147.0 mg/m^2.h) for the *T. latifolia*-planted VSSF CWs were higher than those of the control VSSF CWs (95.2 mg/m^2.h) (Fig. 5.10a). Also, the BOD_5 MRRs for the *T. latifolia* HSSF CWs were higher than those of the control HSSF CWs (Fig. 5.10b and 5.10c). The mean COD MRRs for the T. *latifolia* VSSF CWs (120.3 g/m^2.h) were higher than those of the control wetland systems (72.3 mg/m^2.h). However, the BOD_5 MRRs of the T. *latifolia* VSSF CWs showed a moderate correlation with the incoming loads, with R^2 values of 0.7016 and 0.7421, respectively (Fig.5.10a). Also, the BOD_5 MRRs of the T. *latifolia* HSSF CWs had a very strong linear correlation with the incoming loads (R^2 values of 0.9297), this behaviour corresponds to a first-order kinetics, i.e., the removal rate was proportional to the influent concentration (Calheiros et al., 2007). The control HSSF CWs had a very linear correlation with the incoming loads (R^2 values of 0.0008). This showed a major difference between the planted and control HSSF CWs.

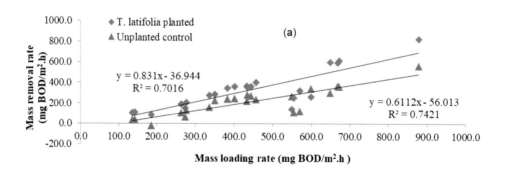

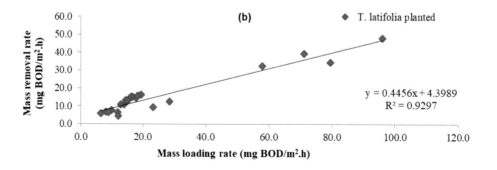

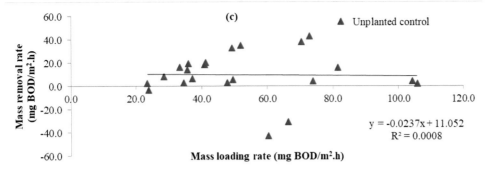

Fig. 5.10. Relationship between BOD₅ mass loading rate and BOD₅ mass removal rate for T. latifolia-planted and control for (a) VSSF CWs; (b) T. latifolia-planted HSSF CW and (c) control HSSF CW.

The COD MRRs of the *T. latifolia* and the control VSSF CWs had a moderate non-linear correlation with the incoming loads (R^2 values of 0.6064 and 0.6151, respectively). While the COD MRRs for the *T. latifolia*-planted and control HSSF CWs were moderately correlated with the incoming loads (R^2 values of 0.8773 and 0.6742, respectively).

There was a high BOD₅ MRR with high MLR in both the *T. latifolia* and control VSSF CW treatment systems and with higher BOD₅ MRRs in the *T. latifolia* HSSF CWs (Fig. 5.10). However, this was not the case with BOD₅ MRRs for the control HSSF CWs, the high MLRs corresponded to lower MRRs. In the case of COD MRRs in both the VSSF CWs and HSSF CWs, a higher COD MLR sometimes corresponded to a higher MRR and a higher COD MLRs corresponded to lower COD MRRs and vice versa (Fig.5.11). This observed behaviour of the planted and control VSSF and HSSF CWs may best be described by the Monod kinetic model (Mitchell and McNevin, 2001), which predicts a first-order behaviour at low concentrations (i.e. pollutant removal rate which increases with increasing pollutant concentration) and zero-order or saturated behaviour at high pollutant concentrations (i.e. a maximum pollutant removal rate) (Mitchell and McNevin, 2001).

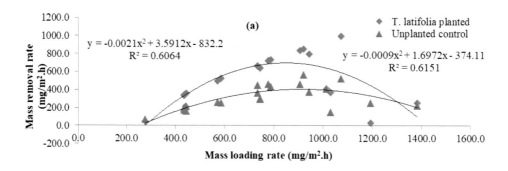

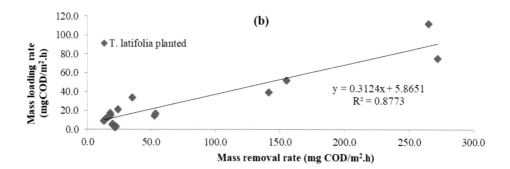

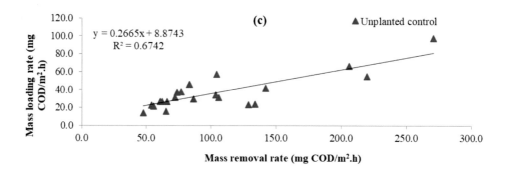

Fig. 5.11. Relationship between COD mass loading rate and COD mass removal rate for T. latifolia-planted and control for (a) VSSF CWs; (b) T. latifolia-planted HSSF CW and (c) control HSSF CW.

5.3.9 HSSF CW (Stage 2) of the hybrid treatment systems

Table 5.3 presents the additional contributions (%) made by the *T. latifolia*-planted HSSF CWs. The highest additional contribution in the HSSF CWs was for the removal of PO_4^{3-}-P, while it had the lowest additional contribution for BOD_5 removal. There was also additional removal of NH_4^+-N (27%) and NO_3^--N (13.9%) in the HSSF CW. The HSSF CW thus produced a better effluent quality than the VSSF CW alone, which was acceptable by the WHO and FEPA (Nigeria).

Table 5.3. Removal by VSSF CW (%) and additional contribution by HSSF CW (%)

	BOD	COD	Nitrate-N	Ammonium-N	Phosphate-P
VSSF CW	76.0	72.7	67.7	70.0	49.4
HSSF CW	9.8	18.8	13.9	27.0	32.2

5.4 Discussion

5.4.1 Efficiency of the VSSF-HSSF CWs in improving the quality of the secondary refinery effluent

This study showed that treatment of secondary refinery wastewater in *T. latifolia*-planted CWs can be further optimized by a hybrid (VSSF + HSSF) CW design. Hybrid constructed wetlands are used to achieve higher treatment efficiencies for suspended solids, nutrients and organic pollutants in treated effluents (Abidi et al., 2009; Chyan et al., 2013; Saeed and Guangzhi, 2012; Zeb et al., 2013) compared with single systems. Hence, the results for pollutant removal in this present study (Table 5.2) confirm the performance of the optimization of the VSSF CWs by providing a better effluent quality and higher removal efficiency that met the discharge limits (Table 5.2) of the WHO and FEPA (Nigeria). Also, Zhang et al. (2015) reviewed the application of CWs for wastewater treatment in tropical and subtropical regions and reported higher removal efficiencies of TSS, COD, NH_4^+-N, NO_3^--N and total nitrogen (TN) in all modes of hybrid CWs as compared to a single CW (VSSF, HSSF and surface flow) systems.

5.4.1.1 Organic matter

The hybrid CWs had higher BOD_5 and COD removal efficiencies of 94 and 88% than the removal efficiencies of the single CW systems (VSSF and HSSF CWs 76 and 73% for BOD_5 and 74 and 65% for COD, respectively). Variability in concentrations of different petroleum refinery wastewaters can affect the biodegradability in CWs (Calheiros et al., 2007). This study showed higher BOD_5 and COD removal rates compared to those reported in Czudar et al. (2011). Like this study, Czudar et al. (2011) worked on post treatment of petrochemical wastewater using *T. latifolia*, *Typha angustifolia* and *Phragmites australis*. However, their system had a lower water temperature of 6.5℃ (operating under temperate conditions) at a water depth of 1 m and a discharge of 200 - 250 m^3/day. The higher rate of BOD$_5$ removal may thus be related to the water temperature (28.4 ± 2.3℃) peculiar to the tropics. Plant growth and microbiological activity (biological processes) are temperature dependent, which increases with increasing temperature resulting into a positive effect on the treatment efficiency (Zhang et al., 2015). Also, the secondary refinery wastewater contained higher biodegradable organics (BOD$_5$/COD ratio of > 0.5). More so, effective removal of BOD$_5$ and COD is attributable to filtration/sedimentation of suspended solids and degradation by microorganisms (Dipu et al., 2010; Seeger et al., 2011; Zhang et al., 2015). Thus, the high COD removal efficiencies in the first stage of the planted hybrid system (VSSF CWs) is related to its ability to reduce the soluble COD by aerobic processes, while the additional removal of suspended solids at the HSSF CWs by sedimentation as well as longer hydraulic retention time (148 h) in the HSSF CWs (Table 5.1) resulted into further removal of COD as suggested by Abidi et al. (2009). In addition, BOD$_5$ and COD are also removed by organic matter mineralization within the wetlands, both by microorganisms and the plant species (Czudar et al., 2011). *T. latifolia* produces oxygen photosynthetically, which is partly transferred into its root zone where its supports aerobic bacteria growing there (Sawaittayothin and Polprasert, 2007). These bacteria use it to subsequently degrade the organic compounds in the secondary refinery wastewater. In addition, the pH of substrate media can influence plant growth (Calheiros et al., 2007). The polished refinery wastewater had a mean pH value of 7.4 (±1.6) which is within the pH range (3.0 to 8.5) for *T. latifolia* development (Calheiros et al., 2007).

5.4.1.2 Nutrients

Nitrogen is one of the pollutants in wastewater that can cause eutrophication and also reduce the level of dissolved oxygen in the receiving water bodies (Saeed and Guangzhi, 2012). Secondary Kaduna refinery treated wastewater is characterized by a low nitrogen content (Table 5.2) as also reported (0.30 - 4.17 mg/L) in Mustapha et al. (2015). Similarly, Seeger et al. (2011) reported an average low NO$_3$$^-$-N influent (0.1 mg/L) concentration for benzene, MTBE and NH$_4$$^+$-N contaminated-groundwater.

The control hybrid CWs had a lower average mass removal efficiency of 45 and 69% NH$_4$$^+$-N and NO$_3$$^-$-N, respectively. This lower performance may be attributed to a limited nitrification capacity of the HSSF CWs (Seeger et al., 2011; Czuda et al., 2011). However, the *T. latifolia* hybrid CWs showed higher average removal efficiencies for NH$_4$$^+$-N and NO$_3$$^-$

-N (84 and 89%, respectively) than the single CW systems, i.e. 70 and 68% for VSSF CW and 43 and 65% for HSSF CW for, NH_4^+-N and NO_3^--N, respectively. The high NH_4^+-N removal rate in the planted systems showed that nitrification was very active (Chung et al., 2008). The *T. latifolia* VSSF and HSSF CWs had an effluent pH ranging from 6.4 to 7.5, this is the optimal pH range for nitrification and denitrification (Travaini-Lima and Sipaúba-Tavares, 2012). Also, the presence of plants in the hybrid CWs reduced the NH_4^+-N and NO_3^--N concentration in the secondary refinery effluent to lower levels than the unplanted CWs (Figs. 5.6a, b and 5.7a, b). This finding of the significance of plant presence in enhancing nitrogen removal in CW agreed with other studies (Chung et al., 2008; Seeger et al., 2011; Chyan et al., 2013; Chapter 4).

The major removal mechanisms for PO_4^{3-}-P removal in the CWs are adsorption, precipitation and plant uptake (Peng et al., 2005) with subsequent harvest (Chyan et al., 2013). The PO_4^{3-}-P removal efficiency in this study for the HSSF CW treatment system had similar removal efficiencies as reported by Chung et al. (2008). The *T. latifolia* hybrid system showed a considerable higher PO_4^{3-}-P removal efficiency (68%) compared to the removal rates of the single CW systems (VSSF: 47% and HSSF: 39%) for this study (Fig. 5.8). This PO_4^{3-}-P removal by the *T. latifolia* hybrid CWs may be due to absorption (Chyan et al., 2013; Sawaittayothin and Polprasert, 2007), resulting in final HSSF effluent concentrations of below 1.5 mg/L when operated at a HRT of 148 h. Besides, *T. latifolia* has an extensive rhizomatous root system (Fig. 5.9) that can remove nitrogen and PO_4^{3-}-P from wastewater (Chung et al., 2008).

Fig. 5.9. Pictorial view of the root system of T. latifolia: initial (A) and harvested at the end of the experiment (B)

5.4.2 Role of *T. latifolia* planted VSSF CWs in the optimization of the treatment system

Plants are an essential part of the design of CWs (Haberl et al., 2003). However, the choice of an appropriate plant species is crucial in enhancing the removal efficiency of a CW (Haberl et al., 2003; Madera-Parra et al., 2015). This study demonstrated a higher performance of *T. latifolia*-planted VSSF CW in the removal of turbidity, BOD_5, COD, NH_4^+-N, NO_3^--N and PO_4^{3-}-P compared to those removed by *C. alternifolius* and *C. dactylon*-planted VSSF CWs (Table 5.5), previously studied by Mustapha et al. (2015) and operating under the same experimental conditions (Table 5.1) with the same wastewater (secondary refinery wastewater). In addition, *T. latifolia* (this study), *C. alternifolius* and *C. dactylon* (Chapter 4) planted VSSF CWs showed higher removal efficiencies and enhanced transformation of the measured pollutants compared to the control VSSF CWs (Table 5.5). This is attributable to the rhizospheric effects (Haberl et al., 2003) as well as an indication for the importance of the plant species selection for pollutant removal by CWs (Seeger et al., 2011). In constructed wetlands, the root zone is the most active zone where physicochemical and biological processes are induced by the interaction of the pollutants with the plants, microorganisms and soil particles (Stottmeister et al., 2003).

Table 5.4. Variations in plant height and number of shoots

Plants	R^2 (height)	R^2 (shoot)
T. latifolia	0.9624	0.9674
C. alternifolius	0.9878	0.9818
C. dactylon	0.9581	0.7421

Furthermore, in the rhizosphere, microbial diversity, density, and activity are more abundant, which promotes increased phytoremediation activity (Hietala and Roane, 2009). The enhanced remediation of contaminants in the rhizosphere is due to high microbial densities and metabolic activities in the rhizosphere, which can be attributable to microbial growth on root exudates and cell debris originating from the plant roots (Weyens et al., 2009).

The *Typha* plant species is an example of a plant with rhizomatic root system with rhizomes and root biomass made up of thick roots, while *C. alternifolius* and *C. dactylon* have a fibril root system with no rhizomes and with fine roots (Wenyin et al., 2007). Wenyin et al. (2007) reported in their study that plants with rhizomatic roots have a higher tolerance to pollution than those with fibril roots. This might explain the higher performance of *T. latifolia* plant species over *C. alternifolius* and *C. dactylon* plant species.

Table 5.5. Comparison between planted T. latifolia, C. alternifolius, C. dactylon and unplanted control VSSF CWs

Constructed wetland	Measured parameters of concern in secondary refinery wastewater (%)						Reference
	Turbidity	BOD	COD	Ammonium	Nitrate	Phosphate	
C. alternifolius	85	68	65	68	58	43	Mustapha et al. (2015)
C. dactylon	82	70	63	49	54	42	Mustapha et al. (2015)
Unplanted	66	47	43	04	16	17	Mustapha et al. (2015)
T. latifolia	89	76	73	70	68	49	This study
Unplanted	64	48	41	20	26	26	This study

5.4.3 Removal efficiency of the HSSF CW treatment systems

The additional removal by the HSSF CWs gave final effluent concentrations that conform very well to the surface water discharge limits of the WHO and FEPA (Nigeria), i.e. 5 NTU for turbidity, 30 mg/L for TSS, 10 mg/L for BOD_5, 40 mg/L for COD and 5 mg/L for PO_4^{3-}-P. Though turbidity, TSS, PO_4^{3-}-P and NH_4^+-N met the effluent discharge limit already at the (first stage) VSSF CW, BOD_5, COD and NO_3^--N concentrations were treated to well below these discharge limit in the HSSF CW. The lower additional BOD_5 removal (9.8%) by the HSSF CW showed that most of the biodegradable organics had already been removed in the VSSF CWs (76%). The effluent from the VSSF CWs also contained more non-biodegradable organics, since a higher COD removal efficiency was observed (Table 5.3). The HSSF CWs showed higher additional removal of NH_4^+-N and PO_4^{3-}-P (Table 5.3), indicating high microbial activity in the HSSF CWs (Sun and Austin, 2007).

5.4.4 Performance of *T. latifolia* in the VSSF-HSSF CWs

This study showed that *T. latifolia* could grow well in the VSSF CWs with a surface area of 0.18 m² and a HSSF CWs with a surface area of 0.77 m² fed with secondary refinery wastewater, without showing any signs of toxicity with the varied influent water qualities. *T. latifolia* grew better in the HSSF CWs than in the VSSF CWs, having both more shoots and higher canopy height in the HSSF CWs (Fig. 5.2). This may be related to the larger surface area (0.77 m²) of the HSSF CWs, which implies that it may have received (123 L) and retained more wastewater. Although, the HSSF CWs received a lower pollutant load than the VSSF CWs that received a higher pollutant load (Table 5.2) and had a lower dissolved oxygen concentration, *T. latifolia* thrived better in the HSSF CWs. Contrary to Calheiros et al. (2009) who attributed better macrophyte growth to a higher organic loading rate in a CW (Table 5.4).

5.4.5 Physicochemical wastewater characteristics

The temperature range recorded for the effluents from the planted and the unplanted VSSF CWs and HSSF CWs (20 – 45 °C) were typical of tropical climates. Such warm temperatures significantly enhance the performance of CWs (Katsenovich et al., 2009). Similarly, the positive effect of high temperatures (20 – 45 °C) on the removal of organic compounds from refinery wastewater was observed in Mustapha et al. (2015) and Saien and Shahrezaei (2012). The average pH value of the VSSF and HSSF CWs effluent from the planted and control were within the range favourable for microbial growth (Kaseva, 2004; Yeh and Wu, 2009). Although, during the course of the study, the influent had a highly varying pH (between 5.9 to 11.3), and the CW treatment systems were able to buffer this effect. The temperature and pH of treated refinery wastewater were reduced to the recommended discharged limits of <40 °C and 6.5 – 9.2, respectively, for the WHO and FEPA, Nigeria (Israel et al., 2008; Nwanyanwu and Abu, 2010).

5.5 Conclusions

The *T. latifolia*-planted hybrid CWs studied demonstrated an effective purification capacity for secondary refinery wastewater. The important findings from this study are as follows:

- The *T. latifolia*-planted hybrid CW had final effluent concentrations that conform very well to the discharge limits of WHO and FEPA (Nigeria), i.e. values of 5 NTU for turbidity, 30 mg/L for TSS, 10 mg/L for BOD_5, 40 mg/L for COD and 5 mg/L for PO_4^{3-}-P.

- In comparison to *C. alternifolius* and *C. dactylon* planted VSSF CWs, *T. latifolia* planted VSSF CW achieved better removal efficiencies for all parameters measured.

- Coupling of a HSSF to VSSF (hybrid CWs) showed very high removal efficiencies of 94% for BOD_5, 88% for COD, 84% for NH_4^+-N, 89% for NO_3^--N, 78% for PO_4^{3-}-P, 85% for TSS and 97% for turbidity.

5.6 Acknowledgement

The authors acknowledge the Netherlands Fellowship Program (NFP) for financial support (NFP-PhD CF7447/2011). Also, the staff of the UNESCO-IHE laboratory and the staff and management of Kaduna Refinery and Petrochemical Company (Kaduna, Nigeria) are acknowledged for their support.

5.7 References

Abidi, S., Kallali, H., Jedidi, N., Bouzaiane, O., & Hassen, A. (2009). Comparative pilot study of the performances of two constructed wetland wastewater treatment hybrid systems. *Desalination, 246*, 370–377.

Abou-Elela, S. I., & Hellal, M. S. (2012). Municipal wastewater treatment using vertical flow constructed wetlands planted with *Canna*, *Phragmites* and *Cyprus*. *Ecological Engineering, 47*, 209-213. Retrieved from http://dx.doi.org/10.1016/j.ecoleng.2012.06.044

Adomokai, R., & Sheate, W. R. (2004). Community participation and environmental decision-making in the Niger Delta. *Environmental Impact Assessment Review, 24*, 495-518. doi:10.1016/j.eiar.2004.01.002

Agbenin, J. O., Danko, M., & Welp, G. (2009). Soil and vegetable compositional relationships of eight potentially toxic metals in urban garden fields from northern

Nigeria. *Journal of Science, Food and Agriculture, 89*(1), 49-54. doi:10.1002/jsfa.3409

Akhavan, S. A., Dejban, G. I., Emani, M., & Nakhoda, A. M. (2008). Isolation and characterization of crude oil degrading *Bacillus* spp. *Iranian Journal of Environmental Health, Science and Engineering, 5*(3), 149-154.

APHA. (2002). *Standard Methods for the Examination of Water and Wastewater* (20th ed.). Baltimore, Maryland, USA: American Public Health Association.

Ávila, C., Garfí, M., & García, J. (2013). Three-stage hybrid constructed wetland system for wastewater treatment and reuse in warm climate regions. *Ecological Engineering*, 43-49. doi:10.1016/j.ecoleng.2013.09.048

Bialowiec, A., Albuquerque, A., & Randerson, P. F. (2014). The influence of evapotranspiration on vertical flow subsurfaceconstructed wetland performance. *Ecological Engineering, 67*, 89-94. doi:http://dx.doi.org/10.1016/j.ecoleng.2014.03.032

Calheiros, C. S., Rangel, A. O., & Castro, P. M. (2009). Treatment of industrial wastewater with two-stage constructed wetlands planted with *Typha latifolia* and *Phragmites australis*. *Bioresources Technology, 100*(13), 3205-3213.

Calheiros, C., Rangel, A., & Castro, P. (2007). Constructed wetland systems vegetated with different plants applied to the treatment of tannery wastewater. *Water Research,* 41,1790 - 1798 (doi:10.1016/j.watres.2007.01.012).

Chazarenc, F., Naylor, S., Comeau, Y., Merlin, G., & Brisson, J. (2010). Modeling the effect of plants and peat on evapotranspiration in constructed wetlands. *International Journal of Chemical Engineering, 2010*, 1-6. doi:10.1155/2010/412734

Chung, A., Wu, Y., Tam, N. F., & Wong, M. H. (2008). Nitrogen and phosphate mass balance in a sub-surface flow constructed wetland for treating municipal wastewater. *Ecological Engineering, 32*, 81-89. doi:10.1016/j.ecoleng.2007.09.007

Chyan, J.-M., Senoro, D.-B., Lin, C.-J., Chen, P.-J., & Chen, M.-L. (2013). A novel biofilm carrier for pollutant removal in a constructed wetland based on waste rubber tire chips. *International Biodeterioration & Biodegradation, In Press*, 1-8 (http://dx.doi.org/10.1016/j.ibiod.2013.04.010).

Czudar, A., Gyulai, I., Kereszturi, P., Csatari, I., Serra-Paka, S., & Lakatos, G. (2011). Removal of organic materials and plant nutrients in a constructed wetlands for petrochemical wastewater treatment. *Studia Universitatis "Vasile Goldiş", Seria Ştiinţele Vieţii, 21*(1), 109-114. Retrieved january 4, 2015, from (www.studiauniversitatis.ro)

Dhulap, V. P., Ghorade, I. B., & Patil, S. S. (2014). Seasonal study and its impact on sewage treatment in the angular horizontal subsurface flow constructed wetlands using aquatic macrophytes. *International Journal of Research in Engineering & Technology, 2*(5), 213-224. doi:ISSN(E): 2321-8843; ISSN(P): 2347-4599

Dipu, S., Anju, A., Kumar, V., & Thanga, S. G. (2010). Phytoremediation of dairy effluent by constructed wetland technology using wetland macrophytes. *Global Journal of Environmental Research, 4*(2).

Haberl, R., Grego, S., Langergraber, G., Kadlec, R. H., Cicalini, A.-R., Dias, S. M., Novais J. M., Aubert, S., Gerth, A., Thomas, H., Hebner, A. (2003). Constructed wetlands for the treatment of organic pollutants. *Journal of Soils and Sediments, 3*(2), 109-124. doi:10.1007/BF02991077

Hagahmed, D. E., Gasmelseed, G. A., & Ahmed, S. E. (2014). Multiple Loops Control of Oil Biodegradation in Constructed Wetlands. *Journal of Applied and Industrial Sciences, 2*(1), 6-13. Retrieved January 4, 2015

Herrera Melián, J., Martín Rodríguez, A. J., Arãna, J., González Díaz, O., & González Henríquez, J. (2010). Hybrid constructed wetlands for wastewater treatment and reuse in the Canary Island. *Ecological Engineering, 36*, 891-899. doi:10.1016/j.ecoleng.2010.03.009

Israel, A. U., Obot, I. B., Umoren, S. A., Mkpenie, V., & Ebong, G. A. (2008). Effluents and Solid Waste Analysis in a Petrochemical Company- A Case Study of Eleme Petrochemical Company Ltd, Port Harcourt, Nigeria. *E-Journal of Chemistry, 5*(1), 74-80. Retrieved from http://www.e-journals.net

Isumonah, V. A. (2015). Minority political mobilization in the struggle for resource control in Nigeria. *The Extractive Industries and Society, 2*, 645-653. doi:10.1016/j.exis.2015.04.011

Ji, G. D., Sun, T. H., & Ni, J. R. (2007). Surface flow constructed wetland for heavy oil-produced water treatment. *Bioresource Technology, 98*(2), 436-441.

Kantawanichkul, S., Kladprasert, S., & Brix, H. (2009). Treatment of high-strength wastewater in tropical vertical flow constructed wetlands planted with *Typha angustifolia* and *Cyperus involucratus*. *Ecological Engineering, 35*, 238-247. doi:10.1016/j.ecoleng.2008.06.002

Kantawanichkul, S., Pilaila, S., Tanapiyawanich, W., & Kamkrua, S. (1999). Wastewater treatment by tropical plants in vertical-flow constructed wetlandss. *Water Science and Technology, 40*(3), 173-178. doi:10.1016/S0273-1223(99)00462-X

Kaseva, M. E. (2004). Performance of a sub-surface flow constructed wetland in polishing pre-treated wastewater - a tropical study. *Water Research, 38*, 681-687. doi:10.1016/j.watres.2003.10.041

Katsenovich, Y. P., Hummel-Batista, A., Ravinet, A. J., & Miller, J. F. (2009). Performance evaluation of constructed wetlands in a tropical region. *Ecological Engineering, 35*, 1529 - 1537. doi:10.1016/j.ecoleng.2009.07.003

Kivaisi, A. K. (2001). The potential for constructed wetlands for wastewater treatment and reuse in developing countries; a review. *Ecological Engineering, 16*, 545-560. doi:PII: S0925-854(00)00113-0

Madera-Parra, C. A., Pena-Salamanca, E. J., Pena, M., Rousseau, D. P., & Lens, P. N. (2015). Phytoremediation of Landfill Leachate with *Colocasia esculenta*, *Gynerum sagittatum* and *Heliconia psittacorum* in Constructed Wetlands. *International Journal of Phytoremediation, 17*, 16-24. doi:10.1080/15226514.2013.828014

Masi, F., & Martinuzzi, N. (2007). Constructed wetlands for the Mediterranean countries: hybrid systems for water reuse and sustainable sanitation. *Desalination, 215*, 44-55. doi:10.1016/j.desal.2006.11.014

Mena, J., Rodriguez, L., Numez, J., Fermandez, F. J., & Villasenor, J. (2008). Design of horizontal and vertical subsurface flow constructed wetlands treating industrial wastewater. *Water Pollution*, 555-557,. doi:10.2495/WP080551

Mitchell, C., & McNevin, D. (2001). Alternative analysis of BOD removal in subsurface flow constructed wetlands employing Monod kinetics. *Water Research*, 1295-1303. doi:PII: S0043-1354(00)00373-0

Murray-Gulde, C., Heatley, J. E., Karanfil, T., Rodgers Jr, J. R., & Myers, J. E. (2003). Performance of a hybrid reverse osmosis-constructed wetland treatment system for brackish oil field produced water. *Water Research, 37*(3), 705-713. doi:PII:S0043-1354(02)00353-6

Mustapha, H. I., van Bruggen, J. J. A., & Lens, P.N. L. (2015). Vertical subsurface flow coonstructed wetlands for polishing secondary Kaduna refinery wastewater in Nigeria. *Ecological Engineering, 85*(http://dx.doi.org/10.1016/j.ecoleng.2015.09.060), 588-595.

NIMET. (2010). Nigeria Meteorological Agency, Nigeria. Kaduna, Kaduna, Nigeria.

Nwanyanwu, C. E., & Abu, G. O. (2010). In vitro effects of petroleum refinery wastewater on dehydrogenase activity in marine bacterial strains. *Ambi-Agua, Taubatè, 5*(2), 21-29. doi:10.4136/ambi-agua.133

Ojumu, T. V., Bello, O. O., Sonibare, J. A., & Solomon, B. O. (2005). Evaluation of microbial systems for bioremediation of petroleum refinery effluents in Nigeria. *African Journal of Biotechnology 4*(1), 31-35. Retrieved from http://www.academicjournals.org/AJB

Peng, J.-F., Wang, B.-Z., & Wang, L. (2005). Multi-stage ponds-wetlands ecosystem for effective wastewater treatment. *Journal of Zhejiang University Science, 6B*(5), 346-352. Retrieved January 5, 2015, from http://www.zju.edu.cn/jzus

Saeed, T., & Guangzhi, S. (2012). A review on nitrogen and organics removal mechanisms in subsurfaceflow constructed wetlands: Dependency on environmental parameters, operating conditions and supporting media. *Journal of Environmental Management, 112*, 429-448. doi:10.1016/j.jenvman.2012.08.011

Saien, J., & Shahrezaei, F. (2012). Organic pollutants removal from petroleum refinery wastewater with nanotitania photocatalyst and UV light emission. *International Journal of Photoenergy, 2012*, 1-5. doi:10.1155/2012/703074

Sawaittayothin, V., & Polpraser, C. (2007). Nitrogen mass balance and microbial analysis of constructed wetlands treating municipal landfill leachate. *Bioresource Technology, 98*, 565-570. doi:10.1016/j.biortech.2006.02.002

Seeger, E. A., Kuschk, P., Fazekas, H., Grathwohl, P., & Kaestner, M. (2011). Bioremediation of benzene-, MTBE- and ammonia-contaminated groundwater with pilot-scale constructed wetlands. *Environmental Pollution, 159*, 3769-3776. doi::10.1016/j.envpol.2011.07.019

Senewo, I. D. (2015). The Ogoni Bill of Rights (OBR): Extent of actualization 25 years later? *The Extractive Industries and Society*(2), 664-670. doi:10.1016/j.exis.2015.06.004

Sepahi, A. A., Golpasha, I. D., Emami, M., & Nakhoda, A. M. (2008). Isolation and characterization of crude oil degrading Bacillus spp. *Iranian Journal Environmental Health, Science and Engineering, 5*(3), 149-154.

Sun, G., & Austin, D. (2007). Completely autotrophic nitrogen-removal over nitrite in lab-scale constructed wetlands: Evidence from a mass balance study. *Chemosphere, 68*, 1120-1128. doi:10.1016/j.chemosphere.2007.01.060

Travaini-Lima, F., & Sipaúba-Tavares, H. L. (2012). Efficiency of a constructed wetland for wastewaters treatment. *Acta Limnologica Brasiliensia, 24*(3), 255-265.

Vymazal, J. (2005). Horizontal sub-surface flow and hybrid constructed wetlands systems for wastewater treatment. *Ecological Engineering, 25*, 478-490. doi:10.1016/j.ecoleng.2005.07.010

Yeh, T. Y., & Wu, C. H. (2009). Pollutant removal within hybrid constructed wetland systems in tropical regions. *Water Science and Technology* , 233-240.

Zapater-Pereyra, M., Ilyas, S., van Bruggen, J., & Lens, P. (2015). Evaluation of the performance and space requirement by three different hybrid constructed wetlands in a stack arrangement. *Ecological Engineering, 82*, 290-300. doi:http://dx.doi.org/10.1016/j.ecoleng.2015.04.097

Zeb, B. S., Mahmood, Q., Jadoon, S., Pervez, A., Irshad, M., Bilal, M., & Bhatti, Z. A. (2013). Combined industrial wastewater treatment in anaerobic bioreactor posttreated

in constructed wetland. *BioMedical Research International, 2013*, 1-8. Retrieved
January 4, 2015, from http://dx.doi.org/10.1155/2013/957853

Zhang, D.-Q., Jinadasa, K., Gersberg, R. M., Liu, Y., Tan, S. K., & Ng, W. J. (2015).
Application of constructed wetlands for wastewater treatment in tropical and
subtropical regions (2000–2013). *Journal of Environmental Sciences, 30*, 30-46.
doi:10.1016/j.jes.2014.10.013

Zurita, F., & White, J. R. (2014). Comparative study of three two-stage hybrid ecological
wastewater systems for producing high nutrient, reclaimed water for irrigationreuse in
developing countries. *Water, 6*, 213-228. doi:10.3390/w6020213

Ch. 6. Fate of heavy metals in vertical subsurface flow constructed wetlands treating secondary treated petroleum refinery wastewater in Kaduna, Nigeria

This chapter has been presented and published as:

Hassana Ibrahim Mustapha., P. N. L. Lens., 2017. Polishing of secondary refinery wastewater for reuse purposes. In: ***Proceedings of the 2017 2nd International Engineering Conference*** (IEC 2017) on green research, innovation and sustainable development in Engineering and Technology: A means to diversification of mono-cultural economies, Minna, Nigeria (17 – 19 October 2017).

Hassana Ibrahim Mustapha., J. J. A. van Bruggen., P. N. L. Lens., 2018. Fate of heavy metals in vertical subsurface flow constructed wetlands treating secondary treated petroleum refinery wastewater in Kaduna, Nigeria. International Journal of Phytoremediation 20(1): 44-56.
Doi: 10.1080/15226514.2017.1337062

Abstract

This study examined the performance of pilot scale vertical subsurface flow constructed wetlands (VSF-CW) planted with three indigenous plants, i.e. *Typha latifolia*, *Cyperus alternifolius* and *Cynodon dactylon* (L.) Pers., in removing heavy metals from secondary treated refinery wastewater under tropical conditions. The total heavy metal concentrations in the influent amounted to 15.6 mg Cu/L, 13.2 mg Cr/L, 12.4 mg Zn/L, 4.4 mg Pb/L 0.93 mg Cd/L and 68.9 mg Fe/L. The *T. latifolia* planted VSF-CW had the best heavy metal removal performance, followed by the *C. alternifolius* planted VSF CW and then the *C. dactylon* planted VSF-CW. The data indicated that Cu, Cr, Zn, Pb, Cd and Fe were accumulated in the plants at all the three VSF-CWs. However, this accumulation of the heavy metals in the plants accounted for only a rather small fraction (0.09 - 16 %) of the overall heavy metal removal by the wetlands. The plant roots accumulated the highest amount of heavy metals, followed by the leaves and then the stem. Cr and Fe were mainly retained in the roots of *T. latifolia*, *C. alternifolius* and *C. dactylon* (TF < 1), meaning that the Cr and Fe were only partially transported to the leaves of these plants. The concentrations of the six heavy metals investigated in the shoots and roots of the three indigenous plants exceeded reported threshold limits. This study showed that VSF-CWs planted with *T. latifolia*, *C. alternifolius* and *C. dactylon* can be used for the large-scale removal of heavy metals from secondary refinery wastewater under tropical conditions.

Keywords: heavy metals, refinery wastewater, constructed wetlands, bioaccumulation, translocation

6.1 Introduction

Heavy metal contamination of our environment is a global issue that has both environmental and health implications. Heavy metals are persistent and accumulate in natural water courses due to domestic and industrial wastewater discharges. This accumulation can lead to detrimental impacts on the receiving ecosystems (Lesage et al. 2007) and humans through the food chain and via drinking water (Cheng et al. 2002). Heavy metals such as cadmium, chromium, cobalt, copper, nickel, mercury, zinc and lead are important pollutants in petroleum refinery wastewater, in addition to total dissolved solids (TDS), organics (dispersed oil, polycyclic aromatic hydrocarbons) and ammonia (Gillespie Jr et al. 2000; Xia et al. 2003; Wuyep et al. 2007; Shpiner et al. 2009). The interest in the use of constructed wetlands (CWs) as an alternative to conventional treatment technologies for oil refinery effluents has grown considerably (Yadav et al. 2012). Heavy metals, unlike organic pollutants, are not degraded by microorganisms. They are removed from wastewater in CWs by immobilizing them in the rhizosphere or by uptake by the microbiota (Khan et al. 2009; Choudhary et al. 2011) and plants (Song et al. 2006). Immobilization in the rhizosphere can occur via adsorption on soil components (Lesage et al. 2007; Stottmeister et al. 2006), precipitation as insoluble salts (Cheng et al. 2002; Cheng 2003) or by filtration (Stottmeister et al. 2006) and sedimentation (Lesage et al. 2007; Choudhary et al. 2011) of suspended particles.

There is a constant search for endogenous (hyper)accumulating plants that can be used in CWs for treating heavy metal contaminated wastewater (Madera-Parra et al. 2015). Liu et al. (2007) demonstrated the uptake and distribution of Cd, Pb and Zn by 19 wetland plant species, and concluded that plants play an important role in the removal of heavy metals from wastewater. They also stated that selection of plant species for use in constructed wetlands will influence the overall heavy metal removal efficiency. Similarly, Madera-Parra et al. (2015) reported that appropriate plant selection was crucial to improve the heavy metal removal efficienccy of CWs treating landfill leachate. Nouri et al. (2009) studied 12 types of plant species for their ability to accumulate heavy metals from the Hame Kasi mine in the northwest Hamadan province (Iran). They observed that plants differ widely in their ability to accumulate heavy metals and that root tissues accumulate higher concentrations of metals than shoots. Their observation was in agreement with that of Kumar et al. (2013), who emphasized that metal accumulation varied within plant tissues.

Phytoremediation, if properly designed and managed, may be used for secondary or tertiary treatment of refinery effluents in the tropics and subtropics to further remove concentrations of heavy metals that could not be removed by conventional methods. Several factors such as type and concentration of heavy metals, plant species, exposure time, temperature, pH and redox potential may affect the removal efficiencies of heavy metals by CWs from wastewater (Cheng 2003; Deng et al. 2004). The aim of this study was to examine the removal of six heavy metals (Cr, Cd, Cu, Zn, Fe and Pb) present in pretreated refinery wastewater of the Kaduna refinery (Nigeria) by vertical subsurface flow constructed wetlands (VSF-CWs) with locally available plant species (*T. latifolia*, *C. alternifolius* and *C. dactylon*). The organic matter and nutrient removal performance of these wetlands was described by Mustapha et al. (2015). This paper evaluates the heavy metal removal efficiency of these VSF-CWs as well as the accumulation and translocation of the heavy metals in the plant parts.

6.2 Materials and methods

6.2.1 Experimental design

The study site was located at the Kaduna Refining and Petrochemical Company (KRPC), Kaduna (Nigeria). The Kaduna state is located in the Northern guinea savannah ecological zone of Nigeria. It lies between Latitude $9\,^0$ N and $12\,^0$ N and Longitude $6\,^0$ E and $9\,^0$ E of the prime meridian. The climatic condition at the study site is categorized by constant dry and wet seasons. The rains begin in April/May and end in October, while the dry season starts late October and stops in March of the subsequent year. About 1211 mm of precipitation falls annually and varies 284 mm between the driest month and the wettest month. Also, the average annual temperature in Kaduna is $25.2\,^0$C with an average of $28.6\,^0$C in April, being the warmest month and at $23.3\,^0$C on average in August being the coldest month of the year. The variation in annual temperature is around $5.3\,^0$C (Climate-data.org, 2016) and a relative humidity varying between 20 and 40% in January and 60 and 80% in July. It has a solar radiation ranging between 20 - 25 Wm^{-2} day^{-1} (Mustapha et al. 2015).

Eight VSF-CWs were constructed in January 2012 for this investigation: six were planted with three different native tropical species, in duplicate and two unplanted controls. They were operated from February to December 2012. The VSF-CWs were situated at the effluent discharge point of the KRPC premises. The native tropical plant species *T. latifolia*, *C. alternifolius* and *C. dactylon* were collected at the Kaduna refinery site for this study. The microcosm VSF-CWs used for this study had dimensions of 0.24 m in radius and 0.55 m in height. Their physicochemical parameters, BOD, COD and nutrient removal capacity have been described by Mustapha et al. (2015). The VSF-CWs had an effective volume of 40 L, designed flow rate of 0.83 L/h, hydraulic loading rate of 4.61 $L/m^2.h$ and a hydraulic retention of time of 48 h. The media used was gravel mixed with coarse sand with a porosity of 0.40. The VSF-CWs were filled with gravel of a particle size of 25 - 36 mm in a layer of 20 cm, starting at the bottom, coarse sand with a particle size of sand of 16 - 25 mm in a layer of 15 cm at the middle and fine sand of 6 -10 mm filled in a layer of 15 cm to the top of the VSF- CWs.

6.2.2 Water and plant sampling

The VSF-CWs were continuously fed by gravity with secondary treated refinery wastewater at a flow rate of 18.96 L/day. 500 ml of influent and effluent samples from each of the eight wetland cells were collected every two weeks for heavy metal determination according to the procedures described in the Standard Methods for the Examination of Water and Wastewater (APHA, 1998). Heavy metal accumulation in the plant tissues was determined after every three months by randomly harvesting 60 g wet weight of each of the plant types from the VSF-CWs.

6.2.3 Analytical methods

Influent and effluents samples (500 ml) collected from eight wetland cells were filtered over 0.45 µm pore size filter paper in order to remove suspended particles. The filtrate were acidified with 0.5 ml of HNO_3 (65%) and HCl acid (2:1 v/v) for extraction and used for the determination of Cr, Cd, Cu, Zn, Fe and Pb concentrations using a Buck 210 Atomic Absorption Spectrophotometer. The determination of heavy metals in the wastewater was carried out every two weeks.

The biomass of *T. latifolia*, *C. alternifolius* and *C. dactylon* collected from the wetlands were thoroughly washed and rinsed under running tap water and then rinsed with deionized water in order to remove any soil particles attached to the plant surface after which they were sorted into leaf, stem and root parts. These were chopped into smaller pieces and weighed. 20 g each of the washed, sorted and chopped plant samples were oven dried at 105 ^{0}C to constant weight for 24 h (APHA, 1998). The dried samples were weighed, grounded into powder and 0.5 g DW (dry weight) of the plant tissues was extracted by acid digestion (HNO_3) for the measurement of Cr, Cd, Cu, Zn, Fe and Pb concentrations using a Buck 210 Atomic Absorption Spectrophotometer as described by Nouri et al. (2009). The readings were taken three times for each of the sample to obtain average values. Tukey (HSD) and Duncan homogeneous analysis was conducted to observe the differences in the heavy metal concentrations among the different treatments.

6.2.4 Treatment efficiency of vertical subsurface flow constructed wetlands

The treatment efficiency of the VSF-CW systems based on mass load was calculated using the equation from Białowiec et al. (2014):

$$R = (Ci) - (Ce)/(Ci) * 100\% \qquad (6.1)$$

where:

Ci is the heavy metal concentration (mg/L) in the influent flowing into the wetlands and Ce is the heavy metal concentration (mg/L) in the effluent flowing out of the wetlands.

The overall contribution by the plants was estimated by multiplying the plant weight by the mean metal concentration in the plants to give the plant metal mass in the plants divided by the total output multiplied by 100%.

6.2.5 Bioaccumulation factor in plant parts

Bioaccumulation factors (BAF) of heavy metals in plants were calculated using the equation from Cheng et al. (2002):

$$BAF = \frac{Metal\ content\ in\ plants}{Metal\ concentrations\ in\ the\ influent} \qquad (6.2)$$

where the metal content in the plant is in mg kg^{-1} DW (dry weight) and the metal concentrations in the influent are in mg L^{-1}.

6.2.6 Translocation factor

The translocation factor (TF) is the ratio of the concentration of a heavy metal in the plant, either stem, leaves or shoots to its concentration in the roots. The TF was calculated according to the equation given by Deng et al. (2004):

$$TF = \frac{C_{Shoots}}{C_{Roots}} \qquad (6.3)$$

where C $_{Shoots}$ is the metal concentration in the plant shoots (mg kg^{-1} DW) and C$_{Roots}$ is the metal concentration in the plant roots (mg kg^{-1} DW).

6.3 Results

6.3.1 Heavy metal removal from the refinery wastewater

The heavy metal concentrations in the influent and effluent samples collected from the eight VSF-CWs are presented in Table 6.1. The variations of the Cr, Cu, Cd, Zn, Fe and Pb concentrations in the effluent samples from the *T. latifolia*, *C. alternifolius* and *C. dactylon* VSF-CWs were significantly lower (p < 0.05) than in the control effluent. The total chromium concentration in the VSF-CW influent was 13.23 (± 0.08) mg/L, the mean influent Cr concentration was 0.60 (± 0.55) mg/L, while the mean effluent Cr concentrations were 0.07 (± 0.057), 0.13 (± 0.122), 0.14 (± 0.142) and 0.41 (± 0.384) mg/L for the *T. latifolia*, *C. alternifolius*, *C dactylon* and the unplanted control VSF-CWs, respectively. The effluent from the planted VSF-CWs demonstrated a significant (p < 0.05) reduction in heavy metal concentrations as compared to the control (non-vegetated VSF-CWs). However, no significant differences were observed in the effluent heavy metal concentrations between the different planted VSF-CWs.

Table 6.1 Influent and effluent concentrations (mg/L) of the VSF-CWs treating heavy metal contaminated secondary refinery wastewater (mean ± SD) and effluent discharge and drinking water quality guidelines in mg/L

Samples	Cr	Cu	Zn	Cd	Fe	Pb
Influent	0.60±0.55 [c]	0.71±0.421 [c]	0.57±0.439 [c]	0.042±0.034 [b]	3.13±1.667 [c]	0.20±0.115 [b]
T. latifolia effluent	0.07±0.057 [a]	0.086±0.053 [a]	0.05±0.022 [a]	0.004±0.007 [a]	0.80±0.581 [a]	0.05±0.029 [a]
C. alternifoilus effluent	0.13±0.122 [a]	0.13±0.092 [a]	0.07±0.022 [a]	0.007±0.011 [a]	1.03±0.645 [a]	0.06±0.038 [a]
C. dactylon effluent	0.14±0.142 [a]	0.15±0.113 [a]	0.08±0.020 [a]	0.008±0.011 [a]	1.13±0.735 [a]	0.08±0.047 [a]
Control effluent	0.41±0.384 [b]	0.40±0.260 [b]	0.31±0.198 [b]	0.032±0.027 [b]	2.38±1.250 [b]	0.16±0.090 [b]
[1]WHO guideline	0.05	1.00	3.00	0.003	0.10	0.01
[2]EU guideline	0.05	2.00	5.00	0.005	-	0.01
FEPA (Nigeria)	0.05	2.00	5.00	0.005	20.0	0.05
[3]Safe limit(mg/L)	-	0.20	2.00	-	5.00	5.00

Values are means ± Standard Deviation of two duplicate VSSF CWs (n=22) taken for a period of 11 months, values on the same columns for each heavy metal with different superscript are significantly different (P≤ 0.05), while those with the same superscript are not significantly different (P ≥ 0.05) as assessed by the Tukey (HSD) and Duncan's Multiple Range Tests.
[1] World Health Organization guideline (Kumar and Puri, 2012).
[2] EU guideline (Mebrahtu and Zerabruk, 2011) ; [3]Safe limits (Korsah, 2011)

The mean removal efficiencies of heavy metals from the refinery wastewater stream based on influent and effluent concentrations (using Equation (6.1)) are presented in Table 6.2. *T. latifolia, C. alternifolius* and *C. dactylon* planted VSF-CW showed a good performance for heavy metal removal. The *T. latifolia* planted VSF-CW showed a metal removal in the order of Cd>Cu>Cr>Zn>Pb>Fe, the *C. alternifolius* VSF-CW metal removal performance was in the order of Cd>Cu>Zn>Cr>Pb>Fe and that of the *C. dactylon* VSF-CW followed the order of Cd>Cu>Cr>Zn>Fe>Pb.

Table 6.2. Heavy metal removal efficiencies (%) of the vertical subsurface flow constructed wetlands investigated

Heavy metal	*Typha latifolia*	*Cyperus alternifolius*	*Cynodon dactylon*	Control
Cr	84.56±7.59[a]	74.49±12.75[b]	71.67±9.94[b]	31.71±10.38[c]
Cu	86.64±5.43[a]	78.01±11.82[b]	75.54±13.48[b]	42.18±17.60[c]
Zn	83.21±13.30[a]	76.06±17.57[ab]	70.67±11.82[b]	34.02±17.55[c]
Cd	95.70±7.35[a]	90.08±11.68[a]	90.08±11.68[a]	35.72±22.59[b]
Fe	74.12±12.06[a]	65.63±12.61[b]	64.86±11.73[b]	22.39±12.30[c]
Pb	77.81±9.69[a]	66.30±11.18[b]	61.44±11.23[b]	22.04±8.24[c]

Values are means ± standard deviation (n = 22) of two replicate wetlands. Values on the same rows with different superscript are significantly different (P≤ 0.05) while those with the same superscript are not significantly different (P ≥ 0.05) as assessed by the LSD, Tukey (HSD) and Duncan's Multiple Range Tests.

Comparing the performance of the three planted VSF-CWs (Table 6.2), the *T. latifolia* VSF-CW had the highest (96%) performance for Cd removal, followed by the *C. alternifolius* VSF-CW (90%) and the *C. dactylon* VSF-CW (90%). The least removed metal was Fe in both the *T. latifolia* (74%) and in the *C. alternifolius* (66%) VSF-CW, while Pb was the least removed metal (61 %) in the *C. dactylon* VSF-CW.

Table 6.3 shows the inflow and outflow rates with the precipitation (rainfall) data as well as the evapotranspiration losses during the 11-month experimental period. The average inflow and outflow rates were 26.57 (±11.8) L/day and 17.56 (±0.56) L/day, respectively (Table 6.3). The monthly rainfall ranged from 0.2 L/day in March to 29.5 L/day in July. The monthly evapotranspiration was in the range of 20.6 to 43.1 L/day during the experimental period. Evapotranspiration was higher in the drier months (October to March). In April/May evapotranspiration was also high. The results (Table 6.3) showed that precipitation influences the rate of evapotranspiration.

Table 6.3. Inflow and outflow rates during the VSF-CW experiment

Operation time (days)[*]	Precipitation, Qp (L/day)	Qi (560 + Qp)	ET (L/day)	Qe (Qi - ET)
28	0.0	20.0	1.4	18.6
56	0.2	18.2	1.4	17.3
84	0.5	19.2	1.2	16.8
112	3.4	21.5	1.0	17.7
140	7.4	26.1	0.9	17.1
168	29.5	48.2	0.7	18.0
196	29.3	48.0	0.7	18.0
224	17.8	35.9	0.7	17.3
252	0.5	18.2	0.9	17.8
280	0.9	19.0	1.1	17.0
308	0.0	18.1	1.2	17.5
Total		26.6		17.6

*Days after start of operation and sampling

6.3.2 Plant height and shoots

The plant heights and the number of plants in each planted cell were recorded immediately after transplant in February 2012 and subsequently at a three months' interval in May, August and November 2012, respectively. The three plant species grew well in the secondary treated refinery effluent fed into the various VSF-CWs. *T. latifolia* showed the most rapid growth, despite being transplanted from outside the refinery premises, followed by *C. dactylon* and *C. alternifolius* (Fig. 6.1A). The average number of *C. alternifolius* shoots in the VSF-CW (Fig. 6.1B) was the highest (225), followed by the *T. latifolia* (105) and *C. dactylon* (78) VSF-CW.

6.3.3 Plant contribution to heavy metal removal in VSF-CWs

Table 6.4 shows the plant contribution to the heavy metal removal and the amount of heavy metal removed by the VSF-CWs as well as the contribution made by the unplanted control VSF-CWs. *T. latifolia* accumulated the most Cr (1.73%) in its tissues compared to *C. alternifolius* (0.46%) and *C. dactylon* (0.34%). The Cr removed by the *T. latifolia* planted VSF-CW was 86.7%, while the *C. alternifolius* VSF-CW and *C. dactylon* VSF-CW showed a removal efficiency of 79.1% and 75.8%, respectively (Table 6.4). The unplanted control VSF-CW reduced Cr, Cu, Zn, Cd, Fe and Pb by 35%, 48%, 48%, 29%, 28% and 25%, respectively, during the 11-month study period (Table 6.4).

6.3.4 Heavy metal concentrations, translocation and bioaccumulation in plant parts

Table 6.5 presents the concentration of heavy metals into the plant tissues, as well as the translocation and bioaccumulation factors of *T. latifolia, C. alternifolius* and *C. dactylon*. The variation in the heavy metal concentrations in the plant tissues differed among the plants. For example, the concentrations of the heavy metals ranged from 2.1 to 5633 mg/Kg, 0.1 to 2832 mg/Kg and 0.9 to 3140 mg/Kg in *T. latifolia, C. alternifolius* and *C. dactylon*, respectively (Table 6.5).

T. latifolia had a BAF that ranged from 0.4 to 82.1, the BAF for *C. alternifolius* ranged from 0.2 to 41.1 and *C. dactylon* ranged from 0.3 to 45.6. The variation in concentration was greater in the belowground tissues than in the aboveground tissues. For instance, the roots of the three plants had a higher BAF than their stems and leaves (Table 6.5).

The translocation ability of *T. latifolia* was in the order of Cu>Zn=Pb>Cr>Cd<Fe (ranging from 7.1 to 0.3). *C. alternifolius* exhibited a translocation ability of Cd=Pb>Zn>Cu>Cr>Fe (ranging from 1.7 to 0.2) and *C. dactylon* showed a decreasing order of Cd>Cu>Cr< Pb=Fe (ranging from 1.5 to 0).

6.4. Discussion

6.4.1 Heavy metal removal from secondary refinery wastewater by VSF-CWs

This study showed that the planted VSF-CWs effectively reduced the concentration of all the heavy metals monitored from the secondary treated refinery wastewater. The heavy metal removal processes were sedimentation and plant uptake. The CWs mass balances suggested that sedimentation accounts for the major fraction of the heavy metal removal (data presented in the appendix A). Comparing the performance of the planted VSF-CWs, the *T. latifolia* VSF-CW was the most effective planted CW studied. Accordingly, *T. latifolia* was used for this study due to its tolerance to wastewater with relatively high and often variable concentrations of organic pollutants and nutrients (Mustapha et al. 2015) and heavy metals (Calheiros et al. 2008). Extending the operation and monitoring of the VSF-CW treatment systems beyond 12 months may be necessary to further reduce the Cr, Cd and Pb concentrations in the secondary treated refinery wastewater. In addition, increasing the HRT, or increasing the size and depth of the VSF-CWs could also further reduce the concentration of the heavy metals in the effluent.

6.4.1.1 Essential metals (Cu, Zn)

Cu and Zn are essential elements for organisms and plants, however, they can become poisonous at excessive concentrations in water bodies (Song et al. 2011). Table 6.1 shows the concentrations in the effluents of the three planted VSF-CWs were below the permissible limits (2.0 mg/L and 5.0, respectively) of the FEPA (1991), WHO (2008), and EU (2014). Although the Cu concentration was low, with continuous application of the secondary refinery wastewater to the VSF-CWs, Cu could potentially accumulate beyond its toxic limits (Korsah, 2011). In addition, the Cu concentrations in the effluents of the planted VSF-CWs were significantly lower ($p < 0.05$) than each other and lower than the non-vegetated VSF-CW (Table 1). The Cu removal efficiency for this study was lower than the 92% reported by Song et al. (2006), who used a CW planted with *Cannas generalis* coupled with a micro-electric field (CW/MEF). The differences in the removal efficiency between the planted VSF-CWs from this study and the CW/MEF (Song et al. 2006) can be attributed to the higher pH in the CW/MEF, which resulted in precipitation and flocculation of the metal ions. Also, the growth of *C. generalis* was enhanced by the micro-electric field and could thus assimilate more metals.

6.4.1.2 Macro elements (Fe)

Iron (Fe) is an important element for all forms of life (Jayaweera et al., 2008), but excessive doses of Fe can lead to hemorrhagic and sloughing of mucosa areas in the stomach (Jayaweera et al., 2008). The VSF-CW systems met the FEPA (Nigeria) permissible limit (20 mg/L). However, the systems exceeded the WHO limit (0.3 mg/L) for drinking water (Table 6.1). Kumari and Tripathi (2015) reported a 48% Fe removal efficiency by *P. australis* from urban sewage mixed with industrial effluents in a batch experiment, which was lower than our findings. However, a combination of *P. australis* and *T. latifolia* enhanced the Fe removal efficiency to 80% with increased retention time (Kumar and Tripathi 2015). In contrast, Jayaweera et al. (2008) reported a much better Fe removal efficiency (97%) for a surface flow CW planted with water hyacinth *(Eichhornia crassipes (Mart.) Solms)* in Fe-rich (9.27 mg/L) industrial wastewater. The Fe removal was mainly due to rhizofiltration and chemical precipitation of Fe_2O_3 and $Fe(OH)_3$, followed by flocculation and sedimentation. The addition of 28 mg/L of total nitrogen and 7.7 mg/L of total phosphorus to the synthetic Fe-rich industrial wastewater may have enhanced the performance of their system.

6.4.1.3 Non-metabolic metals (Cr, Cd, Pb)

This study (Table 6.1) showed that Cd and Pb concentrations in the effluents of the *T. latifolia* planted VSF-CWs were within the WHO, EU and FEPA (Nigeria) permissible limits for the discharge of industrial effluent (Adewuyi and Oluwu, 2012; Taiwo et al. 2012). However, Cr, Cd and Pb concentrations in the effluents of the *C. alternifolius* planted VSF-CW and *C. dactylon* planted VSF-CW as well Cr concentrations in the effluents of the *T. latifolia* planted VSF-CWs exceeded the permissible limits (Table 6.1).

Cr, Cd and Pb are non-essential elements to plants and are considered to cause toxicity at multiple levels (Calheiros et al. 2008). Hence, pollution of water bodies by these elements is a serious environmental problem, threatening not only the aquatic ecosystems but also the human populations through the contamination of drinking water and agricultural crops (Cheng et al. 2002). However, Cr, Cd, and Pb concentrations in the effluents were significantly lower ($p < 0.05$) than the non-vegetated VSF-CW, indicating that the planted VSF-CWs effectively removed Cr, Cd and Pb from the secondary refinery wastewater with removal efficiencies ranging between 75% to 88% for Cr; 90% to 96% for Cd and 61% to 78% for Pb by *T. latifolia*, *C. alternifolius* and *C. dactylon* planted VSF-CWs. The non-vegetated VSF-CWs had removal efficiencies of 31%, 36% and 22% for Cr, Cd and Pb, respectively (Table 6.2). This finding was above the range (50% to 92%) reported by Khan et al. (2009) for the removal of Pb, Cd, and Cr from industrial wastewater with a continuous free flowing wetland planted with eleven different types of aquatic macrophytes, among which *T. latifolia*, *P. australis* and *J. articulatus* were used. This can be attributed to the different type of wastewater, the heavy metal influent concentrations used, the CW design, or the macrophytes used. In contrast, Cheng et al. (2002) achieved a 100% removal efficiency for Cd and Pb, which was higher than the findings of this study. Cheng et al. (2002) used an artificial wastewater polluted by heavy metals in a twin-shaped CW, comprising of a vertical flow (inflow) chamber planted with *C. alternifolius,* followed by a reverse-vertical flow (outflow) chamber planted with *Villarsia exaltata.* This twin/shape design may have enhanced the performance of the treatment system.

Table 6.4. Heavy metal removal by the vertical subsurface flow constructed wetlands based on mass loads and the plant contribution to removal process

Treatment	Cr removal		Cu removal		Zn removal		Cd removal		Fe removal		Pb removal	
	CW (%)	Plant (%)	CW (%)	Plant (%)	CW (%)	Plant (%)	CW (%)	Plant (%)	CW (%)	Plant (%)	CW (%)	Plant (%)
T. latifolia	86.69	1.73	87.81	0.69	79.14	10.39	72.81	16.22	70.68	2.8	76.66	1.25
C. alternifolius	79.05	0.46	82.94	0.09	76.79	9.50	79.77	4.20	61.32	4.82	68.37	0.49
C. dactylon	75.83	0.34	79.85	0.14	77.99	7.47	77.17	4.64	58.14	5.26	63.27	0.38
Control	34.93	-	47.51	-	48.07	-	28.79	-	28.18	-	25.23	-

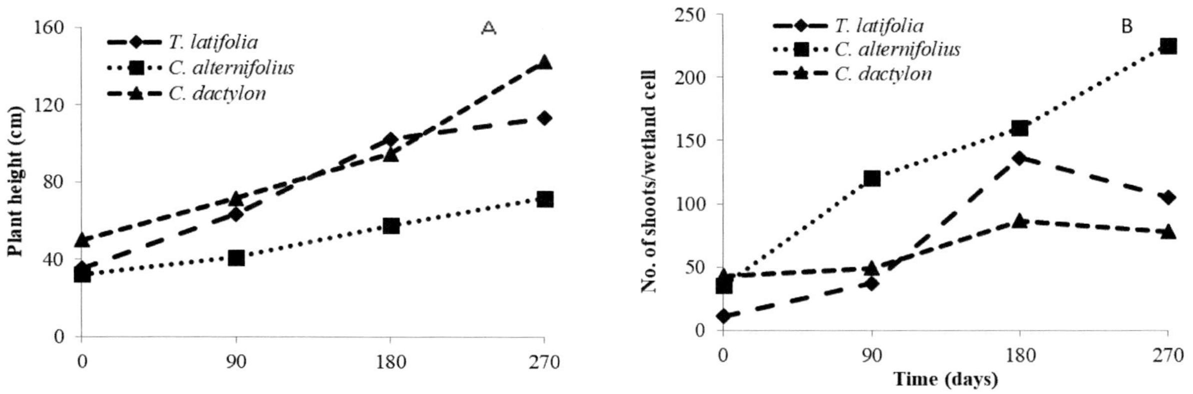

Fig. 6.1. Plant performance in secondary refinery wastewater in terms of plant height (A) and number of shoots per wetland cell (B) as a function of operation time (measured from 2nd February to 7th November, 2012).

Table 6.5. Heavy metal concentrations (mg/Kg), translocation factor (TF) and bioaccumulation factor (BAF) for Typha latifolia, Cyperus alternifolius and Cynodon dactylon growing in the VSF-CW treating secondary refinery wastewater for 308 days. All units are in mg/kg except for TF and BAF that are dimensionless

Plant species	Tissues	Heavy metals concentrations (mg/kg)					
		Cr	Cd	Cu	Zn	Fe	Pb
T. latifolia	Stem[*]	11.4	2.1	6.7	220.5	223.0	16.7
	Leaf[*]	9.7	10.7	34.5	179.0	1330.0	19.2
	Shoot	21.1	12.8	41.2	399.5	1553.0	35.9
	Root[*]	45.1	30.6	5.8	329.7	5653.0	30.0
TF		0.5	0.4	7.1	1.2	0.3	1.2
BAF	Stem	0.9	2.3	0.4	17.8	3.2	3.6
	Leaf	0.7	11.5	2.2	14.5	19.3	4.1
	Root	3.4	32.9	0.4	26.7	82.1	6.5
C. alternifolius	Stem	6.5	0.1	4.9	221.4	49.0	9.3
	Leaf	2.1	15.7	5.9	172.0	447.0	9.5
	Shoot	8.6	15.8	10.8	393.4	496.0	18.8
	Root	17.3	9.1	16.9	419.2	2832.0	11.1
TF		0.5	1.7	0.6	0.9	0.2	1.7
BAF	Stem	0.5	0.1	0.3	17.9	0.7	2.0
	Leaf	0.2	16.9	0.4	13.9	6.5	2.0
	Root	1.3	9.8	1.1	33.9	41.1	2.4
C. dactylon	Stem	5.3	0.9	1.5	157.4	29.0	2.0
	Leaf	4.4	18.9	20.0	26.5	808.0	2.6
	Shoot	9.7	19.2	21.5	183.9	837.0	4.6
	Root	11.5	12.6	16.7	407.2	3140.0	17.4
TF		0.8	1.5	1.3	0.5	0.3	0.3
BAF	Stem	0.4	0.3	0.1	12.7	0.4	0.4
	Leaf	0.3	20.3	1.3	2.1	11.7	0.6
	Root	0.9	13.5	1.1	32.9	45.6	3.7
[**]Phytotoxic range (mg/kg)		5-700	1-10	20-100	100-400	40-500	30-300
[***]Excessive or toxic conc. in mature leaf		5-30	5-30	30-100	100-400	-	30-300
[**]Hyperaccumulation limit (mg/kg)		100	100	1000	10000	10000	1000

[**]Toxic concentrations and hyperaccumulation limits (Korsah 2011). [***] (Kabata-Pendias, 2011)

Keynote: *stem, leaf and roots = (Conc. at end – conc. at start) in each tissue.

6.4.1.4 Unplanted control VSF-CWs

The unplanted control VSF-CW removed 31.7% Cr, 42.2% Cu and 35.7% Cd (Table 6.4) from the secondary treated refinery wastewater fed to it. This might be due to sedimentation of heavy metals in suspended particulate form (Gill et al. 2014) or adsorption of heavy metals on the support media as well as the support system of the PVC pipes installed at the bottom of the VSF-CWs. Although, according to Marchard et al. (2007) in their review of metal and metalloid removal by CWs reported that the medium of unplanted systems will become devoid of binding sites over time, which decreases the capacity of the medium to maintain its metal immobilizing capacity.

6.4.2 Relationship between heavy metal removal by plants and removal by VSF-CWs

In this study, it was observed that the number of shoots and the stem density (biomass production) increased (Fig. 6.1) as the heavy metals accumulated in the plant tissues and the heavy metals removal by the VSF-CW systems increased (Table 6.4). The micronutrients (Zn, Cu), macronutrients (Fe, and Mn) (Deng et al. 2004; Kumar et al. 2013) and the essential nutrients (N, P) (Uchida 2000; Chapter 4) in the treated refinery effluent may have enhanced the plant growth. Therefore, the productivity of the biomass of *T. latifolia*, *C. alternifolius* and *C. dactylon* (Fig. 6.1) may have played an important role in the removal of the heavy metals from the secondary treated refinery wastewater.

However, the contribution of the plants to the removal of heavy metals from the wastewater is only a minor fraction of the overall removal. For example, in the present study, *T. latifolia*, *C. alternifolius* and *C. dactylon* took up 1.73%, 0.46% and 0.34% of Cr, respectively from the refinery wastewater entering the planted VSF-CWs (Table 6.4). Also, *T. latifolia* contributed 1.25% to Pb removal, *C. alternifolius* contributed 0.49% and *C. dactylon* contributed 0.38%. Furthermore, *T. latifolia* gave the highest removal rate for Cu, Zn, Cd and Fe compared with the other two plants used in this study (Table 6.4) as also observed by Zhang et al. (2010) who studied the mangrove plant *Sonneratia apetala Buch-Ham* in a simulated wetland. Furthermore, this research was conducted within the dynamic growth cycle of *T. latifolia*, *C. alternifolius* and *C. dactylon* (perennial herbaceous species). Hence, their performance may be attributed to this dynamic growth cycle. Long-term functioning of constructed wetlands may, nevertheless, increase the heavy metal accumulation in the VSF-CW systems as well as bioaccumulation in the plants and the supporting media beyond the phytotoxic limits (Table 6.5).

Despite the minor repository of *T. latifolia, C. alternifolius* and *C. dactylon* to the overall heavy metal removal efficiency, they contributed to the secondary treated refinery wastewater treatment processes in a number of ways: they favoured the settlement of suspended solids, their rhizosphere provided the substrate and supporting media for the growth of microorganisms, which are the main sites of heavy metals immobilization and uptake in CWs (Maine et al. 2009; Gill et al. 2014). In addition, the plants can carry oxygen from their aerial parts to the roots, creating a proper environment in the rhizosphere for the proliferation of microorganisms and promoting a variety of (bio)chemical reactions which enhanced Cr, Cu, Cd, Zn, Pb and Fe retention by the CW medium (Maine et al. 2009).

Several studies have demonstrated the effectiveness of CWs to remove heavy metals from polluted wastewater (Table 6.6). For instance, the removal rate from the *T. latifolia* VSF-CW, *C. alternifolius* VSF-CW and *C. dactylon* VSF-CW systems (Table 6.6) may indicate that the matrix had a strong adsorption ability (Zhang et al. 2007) and therefore metals are retained in the wetland sediments (Kadlec and Wallace 2009). Furthermore, it was observed that the order of the plant contribution (*T. latifolia* > *C. alternifolius* > *C. dactylon*; Table 6.2) to heavy metal (Cd, Cr, Cu, Fe, Pb and Zn; Table 6.2) removal was consistent with the removal efficiency of the VSF-CWs: T. *latifolia* planted VSF-CW> *C. alternifolius* planted VSF-CW > *C. dactylon* planted VSF-CW.

6.4.3 Bioaccumulation and translocation of metals in wetland plants

6.4.3.1 Bioaccumulation by T. latifolia, C. alternifolius and C. dactylon

None of the plants used in this study accumulated sufficient quantities of heavy metals to qualify them as a hyperaccumulator (Table 6.5). *T. latifolia* had a BAF (Table 6.5) for Cd, Pb and Fe > 1 (BAF= 2.3 to 82.1) suggesting a high accumulation of Cd, Zn and Fe in the soil or CW (Rezvani and Zaefarian 2011). A BAF of < 1 was recorded for Cr and Cu, although the roots and leaves of *T. latifolia* had a BAF > 1 (3.4 and 2.2) for Cr and Cu, respectively. *C. dactylon* is an excluder of Cr and Pb with a BAF < 1. On the contrary, Majid et al. (2014) reported BAF > 1 for Cr (42.9 to 53.8), Cu (375.0 to 1140.0) and for Pb (350.0 to 600.0) for *Typha angustifolia*, indicating *T. angustifolia* is a hyperaccumulator of Cr, Cu and Pb.

6.4.3.2 Heavy metal translocation factor for T. latifolia, C. alternifolius and C. dactylon

Essential metals:

The rate and extent of translocation within each plant species depends on the metal type and plant type (Deng et al. 2004). In this study, the shoots of *T. latifolia* and *C. dactylon* accumulated 41 and 22 mg/kg Cu, respectively. The level of Cu in the shoots of *T. latifolia* in this study was higher than the level of Cu in the shoots of *T. latifolia* reported in Deng et al. (2004). They reported a Cu accumulation of 26 mg/kg in the shoots of *T. latifolia* from a constructed wetland receiving effluents from a Pb/Zn mine. Thus, both *T. latifolia* and *C. dactylon* showed tolerance to Cu (Nouri et al. 2009), which are indeed able to accumulate Cu at concentrations higher than 20 mg/kg (Deng et al. 2004). The Cu concentration in the roots of *T. latifolia* was lower than in the shoots (5.8 mg/kg < 41.2 mg/kg of Cu) indicating a good mobility of Cu from the roots to its shoots, with a very high TF value (TF = 7.1). *C. dactylon* also showed a good mobility for Cu from its roots to the aerial parts. The high translocation of Cu is understandable since Cu is an essential nutrient (Yoon et al. 2006). In contrast, *C. alternifolius* had a TF < 1, meaning that it has a low internal Cu transportation from the roots to its shoots as indicated in Table 6.5.

Zn is also an essential element to plants and is easily taken up by roots (Liu et al. 2007). However, above 230 mg/kg it is considered toxic (Deng et al. 2004). In this study, aboveground tissue had a concentration of 400 mg/kg, 393 mg/kg and 184 mg/kg for *T. latifolia*, *C. alternifolius* and *C. dactylon*, respectively (Table 6.4) and belowground tissue accumulated 330 and 419 mg/kg for *T. latifolia* and *C. alternifolius* and 407 mg/kg for *C. dactylon*. These Zn concentrations for this study are within the levels reported for other emergent plants (Deng et al. 2004). *T. latifolia* showed a TF > 1, whereas *C. alternifolius* and *C. dactylon* had a TF < 1 (0.9 and 0.5, respectively) for Zn. This is an indication of internal transportation of the Zn from the roots to the shoots of *T. latifolia* and a greater availability of this metal in the medium (Nouri et al. 2009) and an exclusion strategy of Zn in *C. alternifolius* and *C. dactylon* (Deng et al. 2004; Bragato et al. 2006).

Macro element:

Fe (5653 mg/kg) was the most accumulated metal in the three plants and the roots of *T. latifolia* contained the highest amount of Fe. *T. latifolia*, *C. alternifolius* and *C. dactylon* had a low mobility of Fe from their roots to the shoots as indicated in Table 6.5. It implies that Fe was largely retained in the roots of these native plants (Denget al. 2004; Jayaweera et al. 2008) and that Fe was slightly soluble in the solution (wastewater) passing through the CW and not easily taken up by the plants (Kabata-Pendias 2011). Hence, *T. latifolia, C. alternifolius* and *C. dactylon* may be classified as Fe excluders (Nouri et al. 2009).

Non-metabolic:

Cr concentrations in the plant tissues ranged from 8.6 - 21.1 mg/kg in the shoots and 11.5 - 45.1 mg/kg in the roots. The TF for Cr in *T. latifolia, C. alternifolius* and *C. dactylon* was < 1 (Table 6.5). The concentrations of Cr in the roots of these plants were higher than in the shoots by a factor more than 1. This indicates a low mobility of Cr towards the aerial parts of the plants. This low translocation from the roots to the shoots may possibly be due to the sequestration of most of the Cr in the vacuoles of the root cells to render it non-toxic, which may be a natural toxicity response of the plants (Marchand et al. 2010). Cr concentrations exceeding 0.5 mg/kg in plant tissues generate toxic effects (Yadav et al. 2011).

Cd is readily taken up by plants and is translocated to different parts of a plant (Liu et al. 2007). The concentration of Cd in the roots of *T. latifolia* was higher than in the shoots (30.6 mg/kg > 12.8 mg/kg of Cr), where a TF < 1 indicates a low mobility of Cd towards the shoots (Nouri et al. 2009). In addition, *T. latifolia* behaves as an excluder, restricting Cd transport to the shoot, thereby keeping the concentration there as low as possible (Bragato et al. 2006). This metal tolerant strategy by *T. latifolia* has also been reported by Deng et al. (2004). *C. alternifolius* had a TF of 1.7 and *C. dactylon* had a TF of 1.5, indicating a high internal heavy metal transportation from the roots to the shoots (TF > 1), as shown in Table 6.5. The average levels of Cd found in the roots of these plants were below the phytotoxic range of 5 - 700 mg Cd /kg.

The translocation ability of *T. latifolia* and *C. alternifolius* for Pb exceeded one (TF = 1.2 and 1.7, respectively), indicating a good internal transportation of Pb from the roots to the shoots of *T. latifolia* and *C. alternifolius*, respectively. On the other hand, *C. dactylon* showed a rather low translocation (TF = 0.3) for Pb, indicating that the capacity of *C. dactylon* to internally transport Pb from its roots to its shoots was limited.

Table 6.6. Overview of heavy metal removal from refinery wastewater by different types of constructed wetlands

Constructed wetland	Wastewater type	Heavy metal	Macrophyte	Removal (%)	Reference
Vertical flow and reverse-vertical flow	Artificial wastewater polluted by heavy metals	Cd, Cu, Pb, Zn, Al, Mn	*C. alternifolius,* *V. exaltata*	100,100,100,100, 100,42 100,100,100,100, 100,42	Cheng et al. 2002
Hybrid: Fresh water wetland	Fresh & blackish water	Cd, Cu, Pb, Zn, Cl	*Schoenoplectus californicus* & *Typha latifolia;*	25,ND,ND,96,NR; 39,89,93,40,12	Kanagy et al., 2008
Saltwater wetland	Saline & hypersaline water		*Spartina alterniflora* & reverse osmosis	100,99,98,99,99; 100,100,99,100,100	
Free water surface	Fe-rich industrial wastewater	Fe	*E. crassipes*	97	Jayaweera et al. 2008
Vertical subsurface flow	Petroleum Refinery	Fe, Cu, Zn	*T. latifolia*	49, 53, 59	Aslam et al., 2010
Horizontal subsurface flow	Diesel refinery wastewater	Cd, Cr, Pb, Zn	*Phragmites australis*	92, 87; 88, 94	Mustapha et al. 2011
Subsurface flow coupled with micro-electric field	Heavy metal containing wastewater	Cd, Cu, Zn	*C. generalis*	69, 92, 51	Song et al. 2011
Vertical subsurface flow	Aqueous solution	Cr, Ni	*Canna indica L.*	98, 96	Yadav et al. 2011
Subsurface flow and free-water surface	Produced oilfield water	Fe, Mn, Ni, Zn	*P. australis;* *T. latifolia*	48-97, 99-100, 64-97, 80-100; NR-89, 88-98, 23-63, 12-84	Horner et al., 2012
Vertical subsurface flow	Aqueous solution	Zn, Ni, Cu,Cr, Co,	*C. indica* *T. angustifolia* *C. alternifolius*	93, NA, 80, NA, 63 99, 76, 68, 66, 55 93, 84, 73, 68, 67	Yadav et al. 2012
Vertical subsurface flow	Petroleum refinery	Cd, Cr, Cu, Zn, Fe, Pb	*C. dactylon* *C. alternifolius* *T. latifolia*	90, 72, 6,71,65,61 90,75,78,76,66,66 96,85,87,83,74,78	This study

ND = below detection limit; NR = no removal; NA = not applicable

6.5 Conclusions

- Vertical subsurface flow constructed wetlands operating under subtropical conditions planted with the locally available plant species *T. latifolia, C. alternifolius* and *C. dactylon* were able to efficiently remove Cd, Cr, Cu, Zn, Fe and Pb from the secondary treated refinery wastewater. Less than 2% of Cr, Cd, Cu, Zn, Fe and Pb removed from the wastewater was stored in the above and belowground biomass of the *T. latifolia, C. alternifolius* and *C. dactylon.*

- *T. latifolia, C. alternifolius* and *C. dactylon* were able to tolerate Cd, Cr, Cu, Zn, Fe and Pb higher than their threshold limits without showing any sign of toxicity. However, the plants did not accumulate sufficient quantities of heavy metals to qualify them as hyperaccumulators.

- The heavy metals were taken up into the root, leaf and stem parts of the plants, with the roots being the most significant. *T. latifolia* had a BAF > 1 (BAF = 2.3 to 82.1) for Cd, Zn and Fe and a BAF < 1 for Cr and Cu. *C. alternifloius* had BAF > 1 for Zn and Pb and BAF < 1 for Cr and Cu. Moreover, *C. alternifloius* and *C. dactylon* are excluders of Cu and Cr and Cr and Pb, respectively with a BAF < 1. The rate and extent of translocation by *T. latifolia, C. alternifolius* and *C. dactylon* depended on the type of heavy metal. *T. latifolia, C. alternifolius* and *C. dactylon* showed translocation factors ranging from 0.3 to 7.1, 0.2 to 1.7 and 0.3 to 1.5, respectively.

6.6 Acknowledgement

The authors acknowledge the management of Kaduna Refinery and Petrochemical Company (Kaduna, Nigeria) for giving the opportunity to conduct this research in their company and the Government of the Netherlands (Netherlands Fellowship Programme, NFP-PhD CF7447/2011) for financial assistance.

6.7 References

Adewuyi GO, Olowu RA. 2012. Assessment of oil and grease, total petroleum hydrocarbons and some heavy metals in surface and groundwater within the vicinity of NNPC oil depot in Aata, Ibadan metropolis, Nigeria. Int J Res Rev App Sci 13(1):166-174. Retrieved May 14, 2016, from

www.arpapress.com/Volumes/Vol13Issue1/IJRRAS_13_1_18.pdf

APHA, AWWA, WEF. 1998. *Standard methods for the examination of water and wastewater* 20th ed. S. C. Lenore, E. G. Arnold, D. E. Andrew, Eds. Baltimore, Maryland: American Public Health Association, American Water Works Association, Water Environment Federation.

Białowiec A, Albuquerque A, Randerson PF. 2014. The influence of evapotranspiration on vertical flow subsurface constructed wetland performance. Ecol Eng 67:89-94. Retrieved from http://dx.doi.org/10.1016/j.ecoleng.2014.03.032

Bragato C, Brix H, Malagoli M. 2006. Accumulation of nutrients and heavy metals in *Phragmites australis (Cav.) Trin. ex Steudel* and *Bolboschoenus maritimus (L.) Palla* in a constructed wetland of the Venice lagoon watershed. Environ Pollut 144(3):967-976. doi:10.1016/j.envpol.2006.01.046

Calheiros CS, Rangel AO, Castro PM. 2008. The Effects of tannery wastewater on the development of different plant species and chromium accumulation in *Phragmites australis*. Achives Environ Contam Toxicol 55(3):404-414. doi:10.1007/s00244-007-9087-0

Cheng S. 2003. Heavy metals in plants and phytoremediation. Environ Sci. Pollut Res 10(5):335-340.

Cheng S, Grosse W, Karrenbrock F, Thoennessen M. 2002. Efficiency of constructed wetlands in decontamination of water polluted by heavy metals. Ecol Eng 18(3):317 - 325. doi:10.1016/S0925-8574(01)00091-X

Choudhary, AK, Kumar S, Sharma C. 2011. Constructed wetlands: an approach for wastewater treatment. Elixir Pollut 37:3666 - 3672. Retrieved from www.elixirpublishers.com

Deng H, Ye ZH, Wong MH. 2004. Accumulation of lead, zinc, copper and cadmium by 12 wetland plant species thriving in metal-contaminated sites in China. Environ Pollut 132(1):29-40. doi:10.1016/j.envpol.2004.03.030

EU. 2014. Eupean Union (Drinking Water) Regulations 2014. Retrieved January 9, 2017, from https://www.fsai.ie/uploadedFiles/Legislation/Food_Legisation_Links/Water/SI122_2 014.pdf : Eupean Union.

FEPA. (1991). Guidelines and standards for environmental pollution control in Nigeria. National Environmental Standards-Parts 2 and 3. Lagos, Lagos, Nigeria: Government Press: Federal Environmental Protection Agency, Nigeria.

Gill LW, Ring P, Higgins NM, Johnston MP. 2014. Accumulation of heavy metals in a constructed wetland treating road runoff. Ecol Eng 70:133-139. doi:10.1016/j.ecoleng.2014.03.056

Gillespie Jr WB, Hawkins WB, Rodgers Jr JH, Cano ML, Dom PB. 2000. Transfers and transformations of Zinc in constructed wetlands mitigation of a refinery effluent. Ecol Eng 14(3):279-292. doi:10.1016/S0925-8574(98)00113-X

Jayaweera MW, Kasturiarachchi JC, Kularatne RK, Wijeyekoon SL. 2008. Contribution of water *hyacinth (Eichhornia crassipes (Mart.) Solms)* grown under different nutrient conditions to Fe-removal mechanisms in constructed wetlands. J Environ Manag 87:450-460. doi:10.1016/j.jenvman.2007.01.013

Kabata-Pendias A. 2011. Trace elements in soils and plants. 4th ed. Boca Raton: CRC Press Taylor & Francis Group. Retrieved January 7, 2017, from http://www.taylorandfrancis.com

Kadlec RH, Scott WD. 2009. Treatment wetlands. 2nd ed. Boca, Raton, FL: CRC Press.

Khan S, Ahmad I, Shah MT, Rehman S, Khaliq A. 2009. Use of constructed wetland for the removal of heavy metals from industrial wastewater. J Environ Manag 50(11):3451–3457. doi:10.1016/j.jenvman.2009.05.026

Korsah PE. 2011. Phytoremediation of irrigation water using *Limnocharis flava, T. latifolia* and *Thalia geniculata* in a constructed wetland. Kwame Nkrumah University of Science and Technology, Department of Theoretical and Applied Biology. Department of Theoretical and Applied Biology.

Kumar M, Pur A. 2012. A review of permissible limits of drinking water. Indian J Occup Environ Med 16(1):40-44. doi:10.4103/0019-5278.99696

Kumar N, Bauddh K, Kumar S, Dwivedi N, Singha DP, Barman S. 2013. Accumulation of metals in weed species grown on the soil contaminated with industrial waste and their phytoremediation potential. Ecol Eng 61 Part A: 491-495. doi:10.1016/j.ecoleng.2013.10.004

Kumari M, Tripathi BD. 2015. Efficiency of *Phragmites australis* and *Typha latifolia* for heavy metal. Ecotoxicol Environm Safety 112:80 - 86. doi:10.1016/j.ecoenv.2014.10.034

Lesage E. 2006. Behaviour of heavy metals in constructed treatment wetlands. Ghent University, Faculty of Bioscience Engineering, Ghent.

Lesage E, Rousseau D, Meers E, Tack F, De Pauw N. 2007. Accumulation of metals in a horizontal subsurface flow constructed wetland treating domestic wastewater in Flanders, Belgium. Sci Total Environ 380:102-115. doi:10.1016/j.scitotenv.2006.10.055

Liu J, Dong Y, Xu H, Wang D, Xu J. 2007. Accumulation of Cd, Pb and Zn by 19 wetland plant species in constructed wetland. Hazard Mat 147(3):947-953. doi:10.1016/j.jhazmat.2007.01.125

Liu J-G, Li G-H, Shao W-C, Xu J-K, Wang D-K. 2010. Variations in uptake and translocation of copper, chromium and nickel among nineteen wetland plant species. Pedosphere 20(1):96-103. doi:10.1016/S1002-0160(09)60288-5

Madera-Parra CA, Pena-Salamanca EJ, Pena M, Rousseau DP, Lens PN. 2015. Phytoremediation of landfill leachate with *Colocasia esculenta*, *Gynerum sagittatum* and *Heliconia psittacorum* in constructed wetlands. Int J Phytorem 17:16-24. doi:10.1080/15226514.2013.828014

Maine MA, Sune N, Hadad HR, Sanchez G, Bonetto C. 2009. Influence of vegetation on the removal of heavy metals and nutrients in a constructed wetland. Environ Manag 90:355-363. doi:10.1016/j.jenvman.2007.10.004

Majid SN, Khwakaram IA, Mam Rasul GA, Ahmed ZH. 2014. Enrichment and translocation factors of some heavy metals in *Typha Angustifolia* and *Phragmites Australis* species growing along Qalyasan Stream in Sulaimani City /IKR. *Journal of Zankoy Sulaimani-, 16 (part A)*(4):93-109. https://www.researchgate.net/publication/269761759

Marchand L, Mench M, Jacob DL, Otte ML. 2010. Metal and metalloid removal in constructed wetlands, with emphasis on the importance of plants and standardized measurements: A review. Environ Pollut 158(12):3447-3461. doi:10.1016/j.envpol.2010.08.018

Masona C, Mapfaire L, Mapurazi S. 2011. Assessment of heavy metal accumulation in wastewater irrigated soil and uptake by maize plants (*Zea Mays L*) at Firle Farm in Harare. J Sustainable Development 4(6):132-137. doi:10.5539/jsd.v4n6p132

Mebrahtu G, Zerabruk S. 2011. Concentration of heavy metals in drinking water from urban areas of the Tigray region, Northern Ethiopia. CNCS, Mckelle University 3, 105-121.

Mustapha HI, van Bruggen JJ, Lens PN. 2015. Vertical subsurface flow constructed wetlands for polishing secondary Kaduna refinery wastewater in Nigeria. Ecol Eng 84:588-595. doi:10.1016/j.ecoleng.2015.09.060

Nouri J, Khorasani N, Lorestani B, Karami M, Hassani AH, Yousefi N. 2009. Accumulation of heavy metals in soil and uptake by plant species with phytoremediation potential. Environ Earth Sci 59(2):315-323. doi:10.1007/s12665-009-0028-2

Qian Y, Gallaghe FJ, Fenga H, Wu M, Zhu Q. 2014. Vanadium uptake and translocation in dominant plant species on an urban coastal brownfield site. Sci Total Environ 476-477(1): 696-704. doi:10.1016/j.scitotenv.2014.01.049

Rezvani M, Zaefarian F. 2011. Bioaccumulation and translocation factors of cadmium and lead in *Aeluropus littoralis*. Austrian J Agric Eng 2(4):114-119.

Shpiner R, Liu G, Stuckey DC. 2009. Treatment of oilfield produced water by waste stabilization ponds: biodegradation of petroleum-derived materials. Bioresour Technol 100(24):6229-6235. doi:10.1016/j.biortech.2009.07.005

Singh D, Gupta R, Tiwari A. 2012. Phytoremediation of lead from wastewater using aquatic plants. J Agric Technol 8(1):1-11. Retrieved from http://www.ijat-aatsea.com

Song X, Yan D, Liu Z, Chen Y, Lu S, Wang D. 2011. Performance of laboratory-scale constructed wetlands coupled with micro-electric field for heavy metal-contaminating wastewater treatment. Ecol Eng 37(12):2061 - 2065. doi::10.1016/j.ecoleng.2011.08.019

Song Z, Zheng Z, Li J, Sun X, Han X, Wang W, Xu M. 2006. Seasonal and annual performance of a full-scale constructed wetlands system for sewage treatment in China. Ecol. Eng 26(3):272-282. doi:10.1016/j.ecoleng.2005.10.008

Stottmeister U, Buddhawong S, Kuschk P, Wiessner A, Mattusch J. 2006. Constructed wetlands and their performance for treatment of water contaminated with arsenic and heavy metals. Soil Water Pollut Monit Protect Remediat 3-23:417-432. Springer Netherlands. doi:10.1007/978-1-4020-4728-2_27

Taiwo A, Olujimi O, Bamgbose O, Arowolo T. 2012. Surface water quality monitoring in Nigeria: situational Aanalysis and future management strategy. D. Voudouris (Ed.). In: Water Quality Monitoring and Assessment. INTECH. pp. 302-319. doi:10.5772/33720

Uchida, R. 2000. Essential nutrients for plant growth. In: Plant nutrient management in Hawaii's Soils, approaches for tropical and subtropical agriculture. J. A. Silva (Ed.). College of Tropical Agriculture and Human Resources, University of Hawaii at Manoa. pp. 31-55. Retrieved January 11, 2017, from http://www.puricare.co.za/UserFiles/File/Essential%20Nutrients%20for%20Plant%20 Growth.pdf

WHO. 2008. Guidelines for drinking water quality. 3rd ed. Geneva: Health Criteria and Supporting Information.

Wuyep PA, Chuma AG, Awodi S, Nok AJ. 2007. Biosorption of Cr, Mn, Fe, Ni, Cu, and Pb metals from petroleum refinery effluent by calcium alginate immobilized mycelia of *polyporus squmosus*. Scientific Res Essay 2(7):217-221. Retrieved from http://www.academicjournals.org/SRE

Xia H, Ke H, Deng Z, Tan P, Liu S. 2003. Ecological effectiveness of vertiver constructed wetlands in treating oil-refined wastewater. Proceedings of the third International conference on Vertiver and exhibition.Guangzhou, China.

Yadav AK, Abbassi R, Kumar N, Satya S, Sreekrishnan TR., Mishra BK. 2012. The removal of heavy metals in wetland microcosms: Effects of bed depth, plant species, and metal mobility. Chem Eng J 211-212:501-507. doi:10.1016/j.cej.2012.09.039

Yadav AK, Kumar N, Sreekrishnan T, Satya S, Bishnoi NR. 2011. Removal of chromium and nickel from aqueous solution in constructed wetland: Mass balance, adsorption–desorption and FTIR study. Chem Eng J 160(1):122-128. doi:10.1016/j.cej.2010.03.019

Yang B, Lan C, Yang C, Liao WB, Chang H, Shu W. 2006. Long-term efficiency and stability of wetlands for treating wastewater of a lead/zinc mine and the concurrent ecosystem development. Environ Pollut 143(3):499-512. doi:10.1016/j.envpol.2005.11.045

Yoon J, Cao X, Zhou Q, Ma QL. 2006. Accumulation of Pb, Cu, and Zn in native plants growing on a contaminated Florida site. Sci Total Environ 368(2-3):456-464. doi:10.1016/j.scitotenv.2006.01.016

Zhang J-E, Liu J-L, Liu, Ounyang Y, Liao B-W, Zhao B.-L. 2010. Removal of nutrients and heavy metals from wastewater with mangrove *Sonneratia apetala Buch-Ham*. Ecol Eng 36(10):807-812. doi:10.1016/j.ecoleng.2010.02.008

Zhang X-B, Liu P, Yang Y-S, Chen W-R. 2007. Phytoremediation of urban wastewater by model wetlands with ornamental hydrophytes. J Environ Sci 19:902-909. Retrieved from www.jesc.ac.an

Ch. 7. Vertical subsurface flow constructed wetlands for the removal of petroleum contaminants from secondary refinery effluent at the Kaduna refining and petrochemical company (Kaduna, Nigeria)

This chapter was submitted for publication as:

Hassana Ibrahim Mustapha., J. J. A van Bruggen., P. N. L. Lens. "Vertical subsurface flow constructed wetlands for the removal of petroleum contaminants from secondary refinery effluent at the Kaduna refining and petrochemical company (Kaduna, Nigeria)" to the Journal of Environmental Science and Pollution Research for publication.

Abstract

Typha latifolia planted vertical subsurface flow constructed wetlands (VSF CWs) and an unplanted microcosm constructed wetland were used for treating secondary refinery wastewater from the Kaduna refining and petrochemical company (KRPC, Nigeria). Cow dung was applied to the planted wetlands at the start of the experiment and after three months to enhance plant growth and petroleum degradation. The *T. latifolia* planted VSF CWs removed 45 - 99 % Total Petroleum Hydrocarbon (TPH), 99 -100 % phenol, 70 - 80% oil and grease, 45 - 91 % chemical oxygen demand (COD) and 46 - 88 % total suspended solids (TSS). The performance of the unplanted control VSF CW achieved lower removal efficiencies: 15 - 58 % TPH, 86 - 91 % phenol, 16 - 44% oil and grease, 24 - 66 % COD and 20 - 55 % TSS. *T. latifolia* plants had a bioaccumulation factor (BAF) > 1 for phenol, total nitrogen (TN) and total phosphate (TP) suggesting a high removal performance for these contaminants and good translocation ability (TF) for TPH, phenol, oil and grease and TN, with the exception of TP which was mainly retained in their roots (BAF = 47). This study showed *T. latifolia* is a good candidate plant to be used in VSF CWs for polishing secondary refinery wastewater in developing countries.

Keywords: Constructed wetlands, Petroleum pollutants, Secondary refinery effluent, Treatment performance, Bioaccumulation, Translocation.

7.1   Introduction

The Nigerian economy is dependent on crude oil and natural gas, but also faces pollution problems from this industry (Mmom and Decker 2010). Increasing attention is paid for developing and implementing innovative technologies for cleaning up the contamination generated by petroleum-based industries (Sathishkumar et al. 2008; Abidi et al. 2009). Wastewaters from the petroleum industry are classified as hazardous industrial wastewater because these contain toxic substances, including phenol, hydrocarbons as well as oil and grease (Shabir et al. 2013; Mustapha et al. 2015) that have serious adverse effects on humans and ecosystems (Clinton et al. 2009). Total petroleum hydrocarbons (TPHs) contain volatile monoaromatic compounds such as benzene, toluene, ethylene and xylene (BTEX), polyaromatic hydrocarbons (PAHs) and aliphatic compounds (Cook and Hesterberg 2013).

Constructed wetlands are an alternative technology that can be used effectively for the treatment of petroleum wastewater (Wallace et al. 2011; Cook and Hesterberg 2013; Mustapha et al. 2015; Mustapha et al. 2018). It can remove multiple contaminants simultaneously by the combined action of the microorganisms growing on the medium and the plant species (Hazra et al. 2011). When cconstructed wetlands are compared with conventional treatment systems, they are low-cost, have low-maintenance, require less energy and they are ecologically friendly (Stottmeister et al. 2003; Chen et al. 2006; Hazra et al. 2011).

Previous studies showed vertical subsurface flow (VSF) CW planted with *Cyperus alternifolius* and *Cynodon dactylon* (Mustapha et al. 2015) and *T. latifolia* (Mustapha et al. 2018) can be used for the treatment of secondary refinery wastewater at Kaduna (Nigeria) for the removal of BOD, COD, ammonium, nitrate, phosphate and total dissolved solids (Muatapha et al. 2015) and heavy metals (cadmium, chromium, lead, zinc, copper and iron) (Mustapha et al. 2018). The removal of specific petroleum compounds such as petroleum hydrocarbons (TPH), phenol, oil and grease was not addressed in these studies. Therefore, the main objective of this study is to evaluate the removal efficiency of VSF CWs planted with *T. latifolia* for the treatment of TPH, phenol, and oil and grease in secondary refinery effluent of the Kaduna refinery that were not removed during the primary and secondary treatment processes. *T. latifolia* was chosen as it gave a better performance over *C. alternifolius* and *C. dactylon* (Mustapha et al. 2018 versus Mustapha et al. 2015). The specific objectives of this study are: (1) to evaluate the treatment performance of the *T. latifolia* planted VSF CW under tropical climatic conditions based on COD, TSS, TN, TP, TPH, phenol, and oil and grease removal from the influent (Kaduna refining effluent) and the efficacy of contaminant removal when compared to an unplanted control, (2) to enhance the treatment performance by adding a slow releasing nutrient (cow dung) to the VSF CWs to aid the degradation rate and (3) to determine the ability of the *T. latifolia* to accumulate and translocate TPH, phenol, oil and grease as well as nutrients (nitrogen and phosphorus) in order to assess it as a candidate for phytoremediation of petroleum-polluted wastewater.

7.2 Materials and methods

7.2.1 Experiment set up

Three VSF CWs were built and set near the effluent discharge channel of the Kaduna refinery (Kaduna, Nigeria with its tropical climate) with cylindrical plastic containers of 0.88 m in height and 0.22 m in radius. Two of the VSF CWs were planted with *T. latifolia* and the third (unplanted) VSF CW was used as control (Fig. 1). The media used was gravel mixed with coarse sand (3 - 35 mm) with a porosity of 40 %. The different particle sizes of the support media were arranged into three layers: 3 – 10 mm, 16 – 25 mm and 25 - 36 mm at 20 cm depth from the top to the bottom of the VSF CWs. The three VSF CWs were fed with tap water for 10 days. Firstly, the secondary treated refinery wastewater was pumped from the effluent discharge channel of the Kaduna refinery into a 5000 litres storage capacity tank. Subsequently, the wastewater was fed into the VSF CWs including the unplanted control starting from day 11. The VSF CWs were operated in a continuous mode. The treatment system had a flow rate of 0.00048 m^3/h and a hydraulic loading rate of 0.0032 m^3/m^2/h.

T. latifolia is a native macrophyte to sub-Saharan Africa and found in the tropical regions (Pardue et al., 2014). Fifty (50) shoots of *T. latifolia* of average height (121 cm) were collected outside the premises of the KRPC (Kaduna, Nigeria). The soil and debris on the plants were washed off. 25 shoots (4.5 kg wet weight) of *T. latifolia* were transplanted into each of the VSF CWs in January 2014 (on day 0). The plants started to wilt on the third day and after day 8, they were completely dried. Therefore, 5 kg of cow dung having 11.86 mg/g total nitrogen and 1.36 mg/g total phosphorus was applied on day 10 and subsequently in February and May 2014 (operation day 44 and day 134, respectively) at the top most layers of the planted VSF CWs as slow releasing nutrient to enhance plant growth and degradation ability of the microbes. In less than ten days (day 18) after the application of the cow dung, new shoots appeared, the VSF CWs were left to mature and acclimatize for three months.

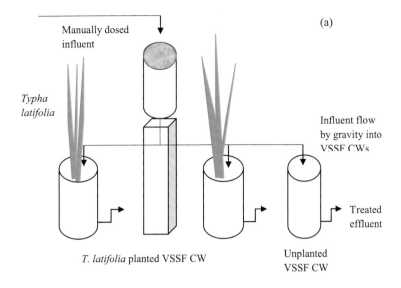

(a)

Manually dosed influent

Typha latifolia

Influent flow by gravity into VSSF CWs

Treated effluent

T. latifolia planted VSSF CW

Unplanted VSSF CW

(b)

Fig. 1. Experimental set up of the VSSF constructed wetlands at Kaduna (Nigeria) used in this study.
(a). Schematic diagram and (b). Photograph of the set up after 150 days after of operation.

7.2.2. Sampling

The plant stem density (new shoots) and plant growth (canopy height) were recorded monthly for each of the VSF CW, the above and belowground biomass were recorded at the start and end of the experiment. Influent and VSF CW effluent samples were taken from operation days 90 to 181 to determine the performance of the *T. latifolia* in reducing the concentrations of carbon oxygen demand (COD), total suspended solids (TSS), total nitrogen (TN), total phosphorus (TP), oil and grease, total petroleum hydrocarbon (TPH) and phenol from secondary refinery effluent using standard methods for the examination of water and wastewater.

New shoots from each of the VSF CWs were measured directly by counting the number of shoots each month. The plant heights were measured with a metric tape to the nearest centimetres from the soil surface every month starting at day 0. At harvest (end of the experiment, day 181), total wet weight, shoots and roots were recorded. Plant tissues were oven dried at 105 ^{0}C for 24 hours, allowed to cool and the dry weight for leaves, stem and roots were determined. The oil and grease, total petroleum hydrocarbon (TPH), phenol, TN and TP concentrations in plant tissues were determined according to standard methods (APHA 2002).

7.2.3 Analysis

At the beginning of the experiment, before transplanting the *T. latifolia* into each of the wetlands, 4.5 kg wet weight of the plants were sorted out into leaves, stem and roots and weighed, and oven dried at 105 ^{0}C for 24 hours. The oven dried samples were digested and analyzed for oil and grease, phenol, TPH, TN and TP. Upon termination of the experiment, the biomass of *T. latifolia* were collected from the wetlands, these were thoroughly washed and rinsed under running tap water and then rinsed with deionized water in order to remove any soil particles attached to the plant surface after which they were sorted into leaf, stem and root parts. These were chopped into smaller pieces and weighed. 20 g each of the washed, sorted and chopped plant samples were oven dried at 105 ^{0}C to constant weight for 24 h (APHA, 1998). The dried samples were weighed, grounded into powder and 0.5 g DW (dry weight) of the plant tissues was extracted by acid digestion (HNO$_3$) for the measurement of O&G, TPH, phenol TN and TP in the tissues of *T. latifolia* were using gravimetric method for O&G described in Pardue et al. (2014) and TPH and phenol analysis were conducted with the method described in Basumatary et al. (2012).

The Hexane Extractable Gravimetric Method, Method 10056 of USEPA (HACH Handbook) was used for O&G and TPH analysis in the plant and wastewater samples. For the determination of phenol, samples were filtered through filters having a pore size of 0.45 μm, the extracted solvent was injected into a gas chromatograph (HP 6890 Powered with HP ChemStation Rev. A 09.01 (1206) software) with capillary column of dimension 30 m (length) x 0.25 mm x 0.25 μm (thickness) and detected by a flame ionization detector (FID). Phenol concentration was determined with a calibration curve made from known phenol standards. The nitrogen and phosphate content in the digested plant tissues, were analyzed by, respectively, APHA 4500-N org (Persulphate) and photometric methods which was determined according to the Standard Methods for the Examination of Water and Wastewater (APHA 2002). All the experiments were duplicated under identical conditions.

Biomass production was determined as described by Hill and Kown (1997) by harvesting the plants at specified time intervals and measuring dry weight production. At end point of the experiment, *T. latifolia* was harvested from each VSF CW, the harvested plants were weighed on site. The harvested plants were sorted out into roots, stems and leaves, and were weighed again.

To determine the reduction of the COD, TSS, TPH, phenol and oil and grease in the secondary refinery effluent, water samples were collected monthly from the inlet and outlet of the three wetlands for both field and laboratory analyses from day 90 till day 180 (April to June 2014). The field parameters analyzed were pH, temperature, total dissolved solids, electrical conductivity and dissolved oxygen. These were measured with hand held equipment (WTW pH 340i, HM TDS - 3 ISO 9001, WTW Cond. 3310 and Oxi 340i, respectively).

COD, TSS, TPH, phenol and oil and grease in the effluent were determined according to standard methods (APHA 2002). The digested plant samples/wastewater samples were extracted with n-hexane, the samples were sonicated in specified pulse mode. Then the extracted solvents were poured into a grade-A 100 mL volumetric flask through a glass funnel that was packed with anhydrous sodium sulphate. However, for TPH, the sample was mixed with silica gel to absorb non-TPH components. The n-hexane was evaporated, the residue left were weighed to determine the concentration of O&G and TPH in mg/L.

7.2.4 Calculations

7.2.4.1 Removal efficiency

The removal of the contaminants (COD, suspended solids, oil and grease, total petroleum hydrocarbon and phenol) from the secondary refinery effluent (influent) and the VSF CW treated effluent were calculated using the following formula from Pardue et al. (2014):

$$Removal\ (\%) = \frac{Concentrations\ inf - Concentrations\ eff}{Concentrations\ inf} * 100 \qquad (7.1)$$

where: Concentrations $_{inf}$ refers to concentrations of contaminant of interest entering into the wetlands (influent) and Concentrations $_{eff}$ refers to the concentrations of contaminants in the effluent that flowed out of the VF CWs.

7.2.4.2 Bioaccumulation of contaminants and translocation factor

The bioaccumulation of contaminants in plant tissues was calculated based on the equation from Shuiping et al. (2002):

$$BAF\ = \frac{Pollutants\ in\ plant\ parts\ (mg/kg\ DW)}{Pollutant\ concentrations\ in\ influent\ (mg/L)} \qquad (7.2)$$

where DW means dry weight.

The translocation factor (TF) was calculated based on the equation from Adewole and Bulu (2012). It is the concentration of contaminants in shoots divided by the concentration in the roots. It is given as:

$$TF\ = \frac{C\ shoots}{C\ roots} \qquad (7.3)$$

where C_{shoots} refers to concentrations of contaminants in shoots and C_{roots} refers to concentrations of contaminants in the roots.

7.3 Results

7.3.1. Physicochemical parameters of VSF CW influent and effluent

Table 7.1 presents the physicochemical properties of the secondary refinery wastewater (influent) and the effluent (treated wastewater) from the VSF CWs. The level of dissolved oxygen increased in the *T. latifolia* planted VSF CW compared to the level in the influent and the unplanted control CW. There was a higher reduction in the total dissolved solids in the *T. latifolia* planted VSF CWs. Variations were observed in the tested parameters with the least variation in pH and the largest variation in the electrical conductivity. The influent pH had a mean value of 7.5 (± 0.1) and the effluent of the *T. latifolia* planted VSF CWs had a mean pH value of 7.4 (± 0.2), whereas the unplanted control CW effluent had mean pH value of 7.5 (± 0.2). The influent had an electrical conductivity of 1.07 (± 0.13) mS/cm and the effluent of the *T. latifolia* planted VSF CW had a mean value of 1.42 (± 0.40) mS/cm. Furthermore, the unplanted control CW had an effluent with a mean conductivity of 9.12 (± 4.17) mS/cm.

Table 7.1. Physicochemical parameters of the influent and effluent of vertical subsurface flow constructed wetlands treating secondary refinery wastewater

Parameter	Sample	N	Mean	Std. Error	Minimum	Maximum
pH	Influent	4	7.5	0.08	7.4	7.7
	T. latifolia - CW	8	7.4	0.21	6.8	8
	Control	4	7.5	0.21	7.2	8.1
Temperature,	Influent	4	32.7	1.54	28.8	36
0 C	T. latifolia - CW	8	28.2	0.79	26.1	29.9
	Control	4	29.8	0.89	27.3	31.5
DO, mg/L	Influent	4	1.2	0.19	0.7	1.6
	T. latifolia - CW	8	3.4	0.52	1.7	4.6
	Control	4	1.7	0.33	1.1	2.5
TDS, mg/L	Influent	4	587	77.15	432	729
	T. latifolia - CW	8	461.1	36.85	330	562
	Control	4	621.8	49.78	479	690
EC, mS/cm	Influent	4	1.07	0.13	0.78	1.34
	T. latifolia - CW	8	1.42	0.4	0.63	2.7
	Control	4	9.12	4.17	0.84	1.03

N=sample size

7.3.2 Removal of TSS and COD

Fig. 7.2 shows the trend in the reduction of the TSS and COD influent concentration for both the *T. latifolia* planted and unplanted control VSF CWs. The TSS and COD concentrations were lower in the *T. latifolia* planted VSF CWs effluent compared to the concentrations in the unplanted control CW (Fig. 7.2). This was the trend throughout the 90 days of monitoring the performance of the VSF CWs. Fig. 7.2 also presents the mean TSS removal from the *T. latifolia* VSF CW and the unplanted control VSF CW. The *T. latifolia* VSF CW had an average TSS removal efficiency between 52 and 78%, while the unplanted control had a TSS removal efficiency between 20 and 55%. The COD removal efficiency of the *T. latifolia* VSF CW ranged between 52 and 83%. The unplanted control had a COD removal efficiency between 24 and 66%.

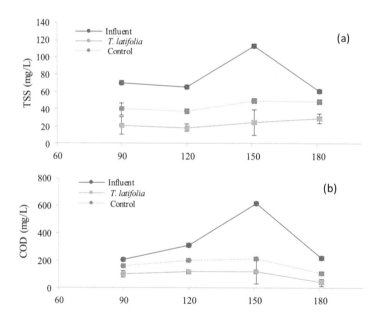

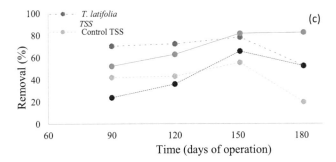

Fig. 7.2. Influent and effluent concentrations from the Typha latifolia planted and unplanted control constructed wetlands: (a) Total suspended solids, (b) COD concentrations and (c) TSS and COD removal efficiency

7.3.3. Petroleum contaminant removal

Fig. 7.3a shows the TPH removal by the *T. latifolia* VSF CW and the unplanted control VSF CW. The TPH concentrations in the *T. latifolia* VSF CW effluent was very low (0.02 to 0.03 mg/L), similarly as in the unplanted control VSF CW (0.03 - 0.04 mg/L) in days 90 and 120. After 150 days, the influent had higher TPH concentrations (6. 2 mg/L) with a better removal observed in the *T. latifolia* VSF CWs (2.0 to 2.3 mg/L) than in the unplanted control CW with an effluent concentration of 5.3 mg/L.

The mean influent phenol concentration was 0.053 µg/L during the studied period (Fig. 7.3b). The effluent phenol concentrations from the *T. latifolia* VSF CWs were below the detection limit (of the GC 0.0001 µg/L), while the unplanted VSF CW had a phenol concentration of 0.001 µg/L. Fig. 7.3b shows the removal efficiencies of the *T. latifolia* and the unplanted control VSF CW. There was almost a complete removal of phenol in the *T. latifolia* planted VSF CWs and equally a high removal in the unplanted VSF CWs which amounted to 99% and 91%, respectively.

Fig. 7.3c presents the oil and grease concentrations in the influent and effluent from the VSF CWs. The influent had concentrations that varied between 1.15 to 6.78 mg/L, the effluent concentrations ranged between 0.28 to 1.92 mg/L and 0.68 to 4.18 mg/L for the *T. latifolia* VSF CWs and the unplanted control VSF CW, respectively. The *T. latifolia* VSF CW had a mean removal efficiency ranging from 72 to 79%, compared to 16 to 44 % by the unplanted control VSF CW (Fig. 7.3c).

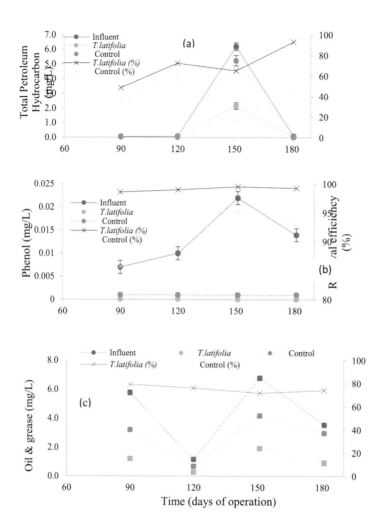

Fig.7.3. *Influent and effluent concentrations and removal efficiency by T. latifolia VSF CW and unplanted control VSF CW: (a) TPH, (b) Phenol and (c) Oil and grease.*

7.3.4 Nutrient removal by VSF CW

There was an increase in the total nitrogen (Fig. 7.4a) and total phosphate (Fig. 7.4b) concentrations in the *T. latifolia* VSF CW effluent. The total phosphate effluent concentration of the *T. latifolia* VSF CW was higher than that of the unplanted control VSF CW and the influent (Fig. 7.4b). However, this stimulated the growth of *T. latifolia* in the VSF CWs (Fig. 7.5).

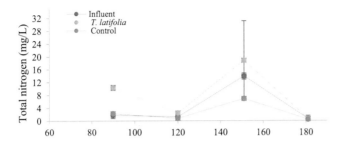

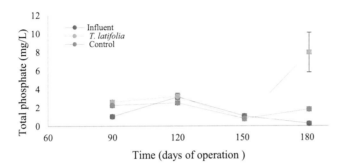

Fig. 7.4. Influent and effluent concentrations (average and standard deviations) from Typha latifolia planted and unplanted control constructed wetlands: (a) total nitrogen and (b) total phosphate

7.3.5. Growth of *Typha latifolia* in the VSF CW

The growth of *T. latifolia* while treating secondary treated refinery wastewater was progressive over time, although it withered for a few days after transplant. The average plant stem density and plant canopy height in the wetlands are shown in Fig. 7.5. Generally, plant establishment showed new growth emerging from the rhizomes after 10 days (Fig. 7.5). There was an over 4-folds more live stem density (plants/0.15 m^2) in the planted wetlands at the end of the six months VSF operation from the 25 shoots that were initially planted (Fig. 7.5). The *T. latifolia* showed a steady increase in canopy height, the highest mean height recorded was 185 cm on day 181 (Fig. 7.5). This steady plant development indicates the tolerance of *T. latifolia* to petroleum contaminants.

The initial 4.5 kg wet weight of the *T. latifolia* planted VSF CW had increased to a mean wet weight of 6.7 kg at the end of the experiment (after 181 days of growth). The roots showed the highest wet weight, followed by the stem and then the leaf. The harvested biomass is presented in Table 7.4. At day 0, 0.61 kg DW of *T. latifolia* was considered for this study. At harvest time, a weight of 1.31 kg DW was harvested, corresponding to a 53% increase of biomass produced.

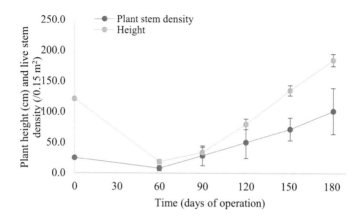

Fig. 7.5. *Evolution of canopy height and stem density of T. latifolia over time in a constructed wetland treating Kaduna refinery wastewater*

7.3.6 Concentration of contaminants in *Typha latifolia* tissue

Fig. 7.6 shows *T. latifolia's* ability to absorb TPH, phenol as well as oil and grease into its tissues while treating secondary refinery wastewater from the Kaduna refinery. The TPH concentrations in the root, stem and leaf of *T. latifolia* were 0.30, 0.24 and 0.94 mg/kg DW, respectively, at the start of the experiment. The mean TPH concentrations at harvest time (day 181) had increased to 2.23 mg/Kg DW root, 1.70 mg/kg DW stem and 2.03 mg/kg DW for the *T. latifolia* VSF CW (Fig. 7.6a). The trend of TPH accumulation was thus root > leaf > stem.

The initial phenol concentration in the roots, stems and leaves was 0.034, 0.037 and 0.039 mg/kg DW (Fig. 7.6b). The end-point mean concentrations were 0.12 mg/kg DW in the root, 0.06 mg/kg DW in the stem and 0.14 mg/kg DW in the leaf. Phenol had thus an accumulation trend of leaf > root > stem for the studied *T. latifolia*.

The initial oil and grease concentrations in the *T. latifolia* was 0.28 mg/kg DW for roots, stems as well as the leaves (Fig. 7.6c). The end-point mean concentrations measured in the different plant parts were 3.56 mg/kg DW in the roots, 3.42 mg/kg DW in the stem tissue and 2.78 mg/kg DW in the leaves of the *T. latifolia*.

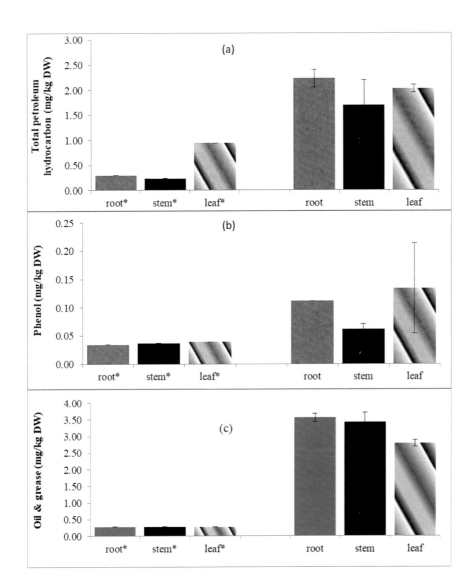

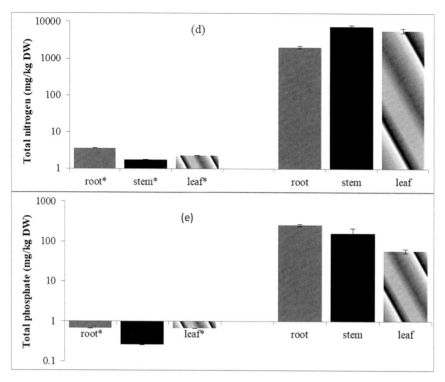

Fig. 7.6. Contaminant concentrations: (a) TPH, (b) Phenol, (c) Oil and grease, (d) total nitrogen and (e) total phosphorus in plant parts at time 0 (*) and at the end of the experiment

7.3.7 Bioaccumulation and translocation factors

Table 7.2 presents the bioaccumulation factor (BAF) and the translocation factor (TF) of TPH, phenol, oil and grease, TN and TP for *T. latifolia* growing in the VSF CW treating secondary refinery effluent. The *T. latifolia* had a bioaccumulation factor between 0.21 and 0.37 for TPH, with a corresponding mean TF of 1.67, indicating the ability of *T. latifolia* to take up TPH into its aboveground part (leaf and stem). The *T. latifolia* also showed a bioaccumulation factor for phenol between 1.0 and 3.5, with a translocation ability of 1.77 for phenol as well as a BAF between 0.24 and 0.33 for oil and grease in it tissues corresponding to a translocation factor of 1.75.

The *T. latifolia* BAF for TN in the plant parts were high with the stem having the highest quantities (425) followed by the leaf (363) and then the root (117). The translocation ability of *T. latifolia* for TN was quite high in the present study. The translocation factor of *T. latifolia* for TN was between 5.67 and 7.42 with a mean TF value of 6.55. The roots of the *T.*

latifolia accumulated more TP than the shoots (Table 7.2). The *T. latifolia* had a mean phosphate BAF of 9.82, 29.36 and 47 for it leaf, stem and roots, respectively, corresponding to a mean TF of 0.86.

Table 7.2. Bioaccumulation factor (BAF) and Translocation factor (TF) for Typha latifolia in vertical subsurface flow constructed wetlands treating petroleum refinery wastewater (mean values (N = 2)

Tissue	Factor	Contaminant				
		TPH	Phenol	Oil and Grease	TN	TP
BAF	Leaf	0.31	1.51	0.26	362.5	9.82
	Stem	0.27	1.15	0.31	425.37	29.36
	Root	0.35	2.11	0.32	116.48	47.00
TF		**1.67**	**1.77**	**1.75**	**6.55**	**0.86**

Table 7.3. Biomass harvested from each VSF CW (kg DW/0.15 m²) treating petroleum refinery wastewater (mean values, N= 2)

Time (days of operation)	Plant tissue	*T. latifolia* cell 1	*T. latifolia* cell 2
0	Root	0.39	0.56
	Stem	0.07	0.09
	Leaf	0.04	0.05
181	Root	0.86	1.24
	Stem	0.15	0.19
	Leaf	0.08	0.10

7.4. Discussion

7.4.1. Performance of constructed wetlands for treatment of petroleum contaminants

This study showed that VSF CWs planted with *T. latifolia* have a high potential to reduce TPH, phenol and oil and grease from secondary refinery effluent (Fig. 7.3). The removal efficiency of oil and grease by the *T. latifolia* VSF CWs in the present study increased over time, with the systems achieving a removal efficiency of 70 - 80 % for the planted system (phytostimulation or rhizodegradation), while only 16 - 44 % for the unplanted VSF CW suggesting the positive effect of the presence of plants on CWs performance (Stefanakis and Tsihrintzis 2012). It was also observed that the removal rates of phenol were higher than those of TPH and O&G (Fig. 7.3) suggesting that phenol was easier to degrade than TPH and O&G (Stefanakis et al. 2016). The factors that can influence the rate of oil pollutant degradation are availability of microorganisms, adequate concentrations of nutrients, oxygen and a pH between 6 and 9 (Das and Chandran 2011). The degradation of organics in the root zone occurs through plant stimulation of microbial activity, by direct uptake by the plants as well as by phytodegradation and volatilization (Dordio and Carvalho 2013). Plants have the ability to stimulate the growth of hydrocarbon-degrading microorganisms in the rhizosphere by releasing root exudates and sloughed-off cells (Jones et al. 2004), thus accelerating the degradation of petroleum contaminants in the secondary refinery effluent.

Wallace et al. (2011) designed an aerated subsurface flow constructed wetland with a surface mulch layer for insulation of the Wyoming Wetland Treatment System (Casper, USA) combined with a free water surface flow CW for groundwater remediation of benzene, BTEX, gasoline-range organics and Fe under cold climate conditions (-35 °C). Their treatment system was most effective in hydrocarbon removal, which removed 100% of these constitutents (non-detectable in the effluent). This was due to the aeration system, multiple years of operation and the CW was a hybrid treatment system with a combined area of 1.3 ha and 90 cm deep gravel beds. These features enhanced both volatilization and aerobic degradation of the hydrocarbons. The present study performed less than the study by Wallace et al. (2011) because the VSFs were not aerated, were single systems (VSF CW only) and operated for a relative short term (181 days).

Basumatary et al. (2012) evaluated the potential of *Cyperus odoratus* L. and *Cyperus laevigatus* for the treatment of TPH in soil and accumulation in plant tissues. Degradation of TPH was greater in the fertilized (73% - 78%) than in unfertilized (43% - 45%) soil. Similarly, in the present study, the application of cow dung to the constructed wetlands enhanced the degradation of TPH, phenol as well as oil and grease. Phenol removal was very high for both the *T. latifolia* VSF CWs and unplanted control CW. Phenol was completely removed in the *T. latifolia* VSF CWs, while the unplanted control CW had a mean removal efficiency of 91% (Fig. 7.2b). The mechanisms for phenol removal can be atttributed to microbial degradation, sorption, plant uptake and volatilization (Ying et al. 2011; Yalcuk 2013).

The application of organic fertilizer (cow dung) may have enhanced oil degradation by stimulation of microorganisms (Fig. 7.3c). Also, Lin and Mendelssohn (1998) applied NPK fertilizer at the loads of 666 kg N ha^{-1} (NH$^+_4$-N), 272 kg P ha^{-1} (P$_2$O$_5$) and 514 kg K ha^{-1} (K$_2$O) to a post-oil spill habitat, 1 and 7 months after transplanting *Spartina alterniflora* and *Spartina patens*, the fertilizer application increased biomass production and regrowth of *S. alterniflora* and *S. patens* within 6 months.

Harvested biomass (Table 7.4) from constructed wetlands is one of the added values of constructed wetlands (Perbangkhem and Polprasert 2010). The mean biomass produced in this study was 1.31 kg for a VSF CW with a surface area 0.15 m^2. *T. latifolia* has high productivity and high biogas potential (Liu et al. 2012). It can be put to various uses such as biofuel production (Liu et al. 2012), food, medicine and paper making (Perbangkhem and Polprasert 2010).

7.4.2. Bioaccumulation and translocation of petroleum contaminants in plant parts

T. latifolia was able to accumulate TPH, phenol as well as oil and grease in its roots, stems and leaves (Fig. 7.6). Over five times more of these contaminants accumulated into its parts compared with the initial concentrations at the start of the experiment. A plant may be considered as a hyperaccumulator if it has a bioaccumulation factor (BAF) greater than 1 (Yoon et al. 2006). In the present study, *T. latifolia* had a BAF < 1 for TPH and oil and grease, indicative of low uptake of hydrophilic compounds of petroleum hydrocarbons (Lotfinasabasl et al. 2013), but a BAF > 1 for phenol indicating the ability of *T. latifolia* to uptake phenol from soil through its roots. Higher values of BAF for phenol suggest that the remediation of phenol by *T. latifolia* was by phytosabilization mechanism (Lotfinasabasl et al. 2013). It also suggests that *T. latifolia* is a phenol hyperaccumulator as proposed by Rezvani and Zaefarian (2011). The BAF values of TPH, phenol and O&G were observed in the order of root > leaf > stem suggesting phytoremediation of potential of *T. latifolia* through phytostabilization in root and phytodegradation in leaf samples (Lotfinasabasl et al. 2013). *T. latifolia* had a good translocation ability for TPH, phenol, oil and grease and TN since the TF > 1 (Deng et al., 2004), with the exception of TP in this study: TP was largely retained in *T. latifolia* roots (Table 7.2). The variation of TF values for *T. latifolia* indicates its potential to treat TPH, phenol and O&G, through translocation into its roots and leaves considering the concentration and type of contaminant (Lotfinasabasl et al. 2013). *T. latifolia* is thus a good candidate plant species for macrophyte based wastewater treatment systems, because of its high productivity, high tolerance, high rate of contaminant accumulation, nutrient assimilation and support to the proliferation of soil microorganisms in the root zone (Mustapha et al., 2018).

The concentration of both the organic contaminants and nutrients were higher in the belowground biomass than the aboveground biomass (Fig 6). *T. latifolia* showed higher accumulation of all the contaminants in its roots compared to the stems and leaves, suggesting that the roots of *T. latifolia* served as a main pathway for organic contaminants and nutrient removal from the *T. latifolia* VSSF CWs. *T. latifolia* showed its potential to readily translocate TPH, phenol, O&G, TN and TP into its other tissues. The mobility of contaminants from the sediment to the plant root (TF > 1) was higher than the mobility within the plant tissues (data not shown). All the contaminants were partly transferred from the sediment to the roots of *T. latifolia*, indicating that uptake and accumulation depended on the pollutant present and its concentration (Lotfinasabasl et al., 2013). It was observed that translocation within the plant tissues depended on the contaminant as well as on the tissue part (Table 7.2). For instance, the transfer of total phosphorus within the *T. latifolia* tissue was higher from root to stem and from root to leaf than from stem to leaf, while TN showed higher translocation from stem to leaf. O&G and phenol was more translocated from the root of *T. latifolia* to it leaf, while TPH and phenol was more transferred from the stem to the leaf of *T. latifolia*. The translocation of phenol from root to leaf (TF = 1.21) is due to mobility within the plant. TPH showed the lowest translocation from root to stem and root to leaf, though it showed a higher translocation from sediment to root than phenol and O&G (Table 7.2). However, O&G and phenol were the most mobile from root/leaf and stem/leaf.

7.4.3 CW effluent quality: TSS and COD

The COD removal efficiency was higher in the *T. latifolia* VSF CW (59 - 91 %) with vegetation than in the unplanted control VSF CW (24 - 66 %). Similarly, the *T. latifolia* VSF CW had a higher TSS removal efficiency than the unplanted control (Fig. 7.2c). The higher removal rate in TSS and COD in the *T. latifolia* VSF CW may be due to the presence of the *T. latifolia* live stem densities (11 to 128 stem 0.18 m^{-2}) compared with the unplanted VSF CW (Fig. 7.5). This finding is similar to the observations made by Debing et al. (2009) in their study of COD, TN and TP removal by poly-culture vegetation structures, using *Typha-Phragmites-Scirpus* (with *T. angustata, P. communis, S. validus* as major species) vegetation, *Typha*-main (with *T. angustata* as major species) vegetation and *Typha*-monoculture vegetation treating artificial sewage, where the poly-culture wetland vegetation of *T. angustata*, *P. communis*, *S. validus*, *Z. latifolia* and *Acorus calamus* with stem densities of 23 stem m^{-2}, 194 stem m^{-2}, 112 stem m^{-2}, 26 stem m^{-2} and 42 stem m^{-2} had the better removal efficiency. Hijosa-Valsero et al. (2012) in their studies also observed that planted systems were more efficient than the unplanted systems, even in winter, in the removal of organic matter and nutrients from urban wastewater. Thus, the better performance of the planted wetland systems is attributed to the presence of plants in the wetlands, especially root penetration may have increased plant root zone aeration, adsorption and uptake of pollutants (Abira 2008).

The *T. latifolia* VSF CW reduced TSS to below the acceptable limits of 30 mg/L (Fig. 7.2a). However, the unplanted control VSF CW did not achieve the FEPA (Nigeria) 30 mg/L TSS discharge limit (Mustapha et al. 2015). Ciria et al. (2005) reported that the mechanism of COD and TSS removal in constructed wetlands is mainly due to physical processes (sedimentation and filtration) instead of biological processes (microbial degradation). This was in confirmation with Ahuja et al. (2011) that COD and TSS removal was related to both physical settling and plant absorption. However, Aslam et al. (2007) stated that COD removal was mainly due to biological degradation and secondarily due to adsorption or absorption to sediments, since their wetlands were barely one year old.

In their study of secondary refinery wastewater at Pakistan, Aslam et al. (2007) reported COD removal efficiencies of 45 - 78% and 33 – 62% for compost and gravel systems with effluent COD concentrations of 55 - 141 mg/L and 89 - 157 mg/L, respectively. They stated that it is difficult to treat secondarily refinery effluent to a COD below 50 mg/L. Indeed, this present study had a mean range of COD removal efficiency of 52 - 83% (Fig. 7.2c) with effluent COD concentrations from 115 mg/L to 38 mg/L.

7.4.4. CW effluent quality: physicochemical parameters

Effluent temperatures ranged from 26 to 30 ^{0}C, which favours the microbial degradation activity in the constructed wetlands (Pan et al. 2012). The effluent temperature was also within the recommended temperature for discharged effluent by the World Health Organization (WHO) and the Federal Environmental Protection Agency (FEPA) of Nigeria. The influent pH was between 7.4 and 7.7 throughout the experiment and the effluent pH ranged from 6.8 to 8.1. A similar pH and temperature were reported for secondary refinery wastewater by Mustapha et al. (2015). The pH value of this present study favoured the uptake of P (Table 7.2; Ying et al. 2011), corresponding to improved growth of *T. latifolia* (Fig. 7.5) and degradation of the contaminants (Fig. 7.2 and Fig. 7.3). The pH was within the WHO (6.5 - 9.2) and FEPA (6.0 - 9.0) recommended limits (Israel et al. 2008).

The *T. latifolia* VSF CWs effluent had a higher increase in DO concentration than in the unplanted control VSF CW. This increase may be due to atmospheric oxygen diffusion inside the matrix pores (Qianxin and Mendelssohn2009) and transfer of oxygen in the VSF CWs as the wastewater flows down the system and also through the plant roots.

The electrical conductivity (EC) of the influent was lower than the effluent except at day 181. Conductivity generally increased across the planted system (Fig. 7.4) which contrasts the findings in Mustapha et al. (2015). This could be due to the addition of cow dung to the *T. latifolia* planted VSF CWs (Schaafsma et al. 2000).

7.5. Conclusions

Two microcosm VSF CWs planted with *T. latifolia* and one unplanted control were operated under tropical climatic conditions (Nigeria) for over 150 days to examine the potential of VSF CWs to remove TPH, phenol as well as oil and grease from secondary refinery wastewater. Cow dung was applied to the *T. latifolia* VSF CW to enhance the growth and degradation processes. The VSF CWs showed effective removal of these contaminants (45 - 99 % TPH; 99 - 100 % phenol and 70 - 80 % oil and grease). *T. latifolia* was able to accumulate TPH, phenol, oil and grease, TN and TP into its plant tissue and showed a high translocation ability for TPH, phenol, oil and grease and TN, with the exception of TP that was mainly retained in the *T. latifolia* roots. The variations in TF values showed that *T. latifolia* is capable of removing petroleum contaminants through its roots, leaves and stem. *T. latifolia* is thus a good candidate plant species that can be used for further polishing secondary refinery wastewater and other types of wastewaters, especially for Nigeria or other developing countries that require low cost and low maintenance wastewater treatment systems.

7.6 Acknowledgement

The authors acknowledge the management of the Kaduna Refinery and Petrochemical Company (Kaduna, Nigeria) for giving the opportunity to conduct this research in their company. We also thank the Government of the Netherlands for their financial assistance (the NUFFIC program) (NFP-PhD CF7447/2011) and the TETFUND (Tertiary Education Trust Fund) for staff training through the Federal University of Technology, Minna (Nigeria).

7.7 References

Abidi S, Kallali H, Jedidi N, Bouzaiane O, Hassen A (2009) Comparative pilot study of the performances of two constructed wetland wastewater treatment hybrid systems. Desalination 246: 370–377.

Adewole MB, Bulu YI (2012) Influence of different organic - based fertilizers on the phytoremediating potential of *Calopogonium mucunoides* Desv. from crude oil polluted soils. Bioremed Biodegrad 3(4): 1-6. http://dx.doi.org/10.4172/2155-6199.1000144

Ahuja S, Sharma HK, Bhasin SK, Dogra P, Khatri S (2011) Removal of contaminants using plants: A review. Current trends in Biotechnol Chem Res 1(1), 11-21.

Alobaidy AH, Al-Sameraiy MA, Kadhem AJ, Abdul Majeed A (2010) Evaluation of treated municipal wastewater quality for irrigation. J Environ Protect 1:216-225. doi:10.4236/jep.2010.13026

APHA (2002) Standard methods for the examination of water and wastewater. 20th ed. Baltimore, Maryland, USA.APHA.

Arivoli A, Mohanraj R (2013) Efficacy of *Typha angustifolia* based vertical flow constructed wetland system in pollutant reductionn of domestic wastewater. Int J Environ Sci 3(5):1497-1508.

Aslam MM, Malik M, Baig M, Qazi I, Iqbal J (2007) Treatment performances of compost-based and gravel-based vertical flow wetlands operated identically for refinery wastewater treatment in Pakistan. Ecol Eng 30: 34 -42. doi:10.1016/j.ecoleng.2007.01.002

Basumatary B, Saikia R, Bordoloi S, Das HC, Sarma, HP (2012) Assessment of potential plant species for phytoremediation of hydrocarbon-contaminated areas of upper Assam, India. J. Chem Technol Biotechnol 87:1329-1334. 10.1002/jctb.3773.

Camacho VJ, Rodrigo MA (2017) The salinity effects on the performance of constructed wetland-microbial fuel cell. Ecol Eng 107: 1-7. https://doi.org/10.1016/j.ecoleng.2017.06.056

Chen TY, Kao CM, Yeh TY, Chien HY, Chao AC (2006) Application of a constructed wetland for industrial wastewater treatment: A pilot-scale study. Chemosphere 64:497-502. doi:10.1016/j.chemosphere.2005.11.069

Cheng S, Grosse W, Kerrenbrock F, Thoennessen M (2002) Efficiency of constructed wetlands in decontamination of water polluted by heavy metals. Ecol Eng 18(3):317-325. doi: 10.1016/S0925-8574(01)00091-X

Ciria MP, Solano ML, Soriano P (2005) Role of macrophyte *Typha latifolia* in a constructed wetland for wastewater treatment and assessment of its potential as a biomass fuel. Biosyst Eng 92(4):535-544. doi:10.1016/j.biosystemseng.2005.08.007

Clinton H, Ujagwung Snr GU, Horsefall Jnr M (2009) Evaluation of total hydrocarbon levels in some aquatic media in an oil polluted mangrove wetland in the Niger Delta. Appl Ecol Environ Res *7*(2):111-120. doi:1589 1623

Cook RL, Hesterberg D (2013) Comparisonof trees and grasses for rhizoremediation of petroleum hydrocarbons. Int J Phytoremed 15:844-860.

Das N, Chandran P (2011) Review article. Microbial degradation of petroleum hydrocarbon contaminants: an overview. Biotechnol Res Int 2011:1-13. doi:10.4061/2011/941810

Debing J, Lianbi Z, Xiaosong Y, Jianming H, Mengbin Z, Yuzhong W (2009) COD, TN and TP removal of *Typha* wetland vegetation of different structures. Polish J Environ Stud 18(2):183-190.

Deng H, Ye ZH, Wong MH (2004) Accumulation of lead, zinc, copper and cadmium by 12 wetland plant species thriving in metal-contaminated sites in China. Environ Pollut 132:29-40. doi:10.1016/j.envpol.2004.03.030

Dordio A, Carvalho A (2013) Organic xenobiotics removal in constructed wetlands, with emphasis on the importance of the support matrix. J Hazard Mater 272-292. doi:10.1016/j.jhazmat.2013.03.008

Hazra M, Avishek K, Pathak G (2011) Developing an artificial wetland system for wastewater treatment: A designing perspective. Int J Environ Protect 1(1):8-18. www.ijep.org

Hijosa-Valsero M, Sidrach-Cardona R, Bécares E (2012) Comparison of interannual removal variation of various constructed wetland types. Sci Total Environ 430:174-183. doi:10.1016/j.scitotenv.2012.04.072

Hill DT, Kown SR (1997) Ammonia effects on the biomass production of five constructed wetland plant species. Bioresourc Technol 62(3):109-113. https://doi.org/10.1016/S0960-8524(97)00085-0

Israel AU, Obot IB, Umoren SA, Mkepenie V, Ebong GA (2008) Effluents and solid waste analysis in a petrochemical company- a case study of Eleme petrochemical company Ltd, Port Harcourt, Nigeria. E-J Chem 5(1):74-80. http://www.e-journals.net

Jones RK, Sun WH, Tang C-S, Robert FM (2004) Phytoremediation of petroleum hydrocarbons in tropical coastal soils II. Microbial response to plant roots and contaminant. Environ Sci Pollut Res 11(5):340-346. doi:10.1065/espr2004.05.199.2

Lin Q, Mendelssohn IA (1998) The combined effects of phytoremediation and biostimulation in enhancing habitat restoration and oil degradation of petroleum contaminated wetlands. Ecol Eng 10:263-274. doi:S0925-8574(98)00015-9

Liu D, Wu X, Chang J, Min Y, Ge Y, Shi Y, Xue H, Peng C, Wu J (2012) Constructed wetlands as biofuel productionsystems. Nature Climate Change 2(3):190-194. doi:10.1013/NCLIMATE1370

Lotfinasabasl S, Gunale VR, Rajurkar NS (2013) Petroleum hydrocarbons pollution in soil and its bioaccumulation in mangrove species, *Avicennia marina* from Alibaug Mangrove Ecosystem, Maharashtra, India. Int J Advan Res Technol 2(2):1-7.

Lu M, Zhang Z, Yu W, Zhu W (2009) Biological treatment of oilfield-produced water: a field pilot study. Int Biodeter Biodegrad 63:316-321. doi:10.1016/j.ibiod.2008.09.009.

Mmom PC, Decker T (2010) Assessing the effectiveness of land farming in the remediation of hydrocarbon polluted soils in the Niger Delta, Nigeria. Res J Appl Sci Eng Technol 2(7): 654-660. doi:2040-7467

Mustapha HI, van Bruggen JJ, Lens PN (2018) Fate of heavy metals in vertical subsurface flow constructed wetlands treating secondary treated petroleum refinery wastewater in Kaduna, Nigeria. Int J Phytoremed 20(1):44-53. doi:10.1080/15226514.2017.1337062

Mustapha H.I, van Bruggen JJ, Lens PN (2015) Vertical subsurface flow constructed wetlands for polishing secondary Kaduna refinery wastewater in Nigeria. Ecol Eng 84:588-595.

Pan J, Zhang H, Li W, Ke F(2012) Full-scale experiment on domestic wastewater treatment by combining artificial aeration vertical- and horizontal-flow constructed wetlands system. Water Air Soil Pollut 223(9):5673-5683. doi:10.1007/s11270-012-1306-2

Pardue MJ, Castle JW, Rodgers Jr JH, Huddleston III GM (2014) Treatment of oil and grease in produced water by a pilot-scale constructed wetland system using biogeochemical processes. Chemosphere 103:67-73. http://dx.doi.org/10.1016/j.chemosphere.2013.11.027

Perbangkhem T, Polprasert C (2010) Biomass production of papyrus (*Cyperus papyrus*) in constructed wetland treating low-strength domestic wastewater. Bioresourc Technol 101:833-835. doi:10.101+/j.biortech.2009.08.062

Qianxin L, Mendelssohn IA (2009) Potential of restoration and phytoremediation with *Jincus roemerianus* for diesel - contaminated coastal wetlands. Ecol Eng 35:85-91. doi:10.1016/j.ecoleng.2008.09.010

Rezvani M, Zaefarian F (2011) Bioaccumulation and translocation factors of cadmium and lead in *Aeluropus littoralis*. Australian J Agric Eng 2(4):114-119. doi:1836-9448

Sathishkumar M, Binupriya AR, Baik S-H, Yun S-E (2008) Biodegradation of crude oil by individual bacterial strains and a mixed bacterial consortium isolated from hydrocarbon contaminated areas. Clean 36(1):92-96.

Schaafsma JA, Baldwin AH, Streb CA (2000) An evaluation of a constructed wetland to treat wastewater from diary farm in Maryland, USA. Ecol Eng 14:199-206. doi:S0925-8574(99)00029-4

Shabir G, Afzal M, Tahseen R, Iqbal S, Khan QM, Khalid ZM (2013) Treatment of oil refinery wastewater using pilot scale fed batch reactor followed by coagulation and sand filtration. American J Environ Protect 1(1):10-13. doi:10.1269/env-1-2

Sharp JL (2002) Managing cattail *(Typha latifolia)* growth in wetland systems. Dissertation, University of North Texas.

Shpiner R, Liu G, Stuckey DC (2009) Treatment of oilfield produced water by waste stabilization ponds: biodegradation of petroleum-derived materials. Bioresour Technol 100:6229-6235. doi:10.1016/j.biortech.2009.07.005

Shutes RB (2001) Artificial wetlands and water quality improvement. Environ Int 26:441-447.

Song Z, Zheng Z, Li J, Sun X, Han X, Wang W, Xu M (2006) Seasonal and annual performance of a full-scale constructed wetlands system for sewage treatment in China. Ecol Eng 26:272-282. doi:10.1016/j.ecoleng.2005.10.008

Stefanakis AI, Tsihrintzis VA (2012) Effects of loading, resting period, temperature, porous media, vegetation and aeration on performance of pilot-scale vertical flow constructed wetlands. Chem Eng J 181-182:416-430. doi:10.1016/j.cej.2011.11.108

Stottmeister U, Wießner A, Kuschk P, Kappelmeyer M, Kästner M, Bederski O, Müller RA, Moormann H (2003) Effects of plants and microorganisms in constructed wetlands for wastewater treatment. Biotechnol Advan, 22:93-117. doi:10.1016/j.biotechadv.2003.08.010

Suhendrayatna M A R, Fajriana Y, Elvitriana (2012) Removal of municipal wastewater BOD, COD, and TSS by phyto-reduction: a laboratory–scale comparison of aquatic plants at different species *Typha latifolia* and *Saccharum spontaneum*. Int J Eng Innovative Technol 1-5.

Wallace S, Schmidt M, Larson E (2011) Long term hydrocarbon removal using treatment wetlands. SPE Annual Technical Conference and Exhibition (pp. 1-10). Denver,

Colarando. Society of Petroleum Engineers.
http://naturallywallace.com/docs/108_SPE%20145797%20Wallace%202011.pdf

Yang L, Hu CC (2005) Treatments of oil-refinery and steel-mill wastewaters by mesocosm constructed wetland systems. Water Sci Technol 51(9):157-164.

Ying X, Dongmei G, Judong L, Zhenyu W (2011) Plant-microbe interactions to improve crude oil degradation. Energy Procedia 5:844-848. doi:10.1016/j.egypro.2011.03.149

Yoon J, Cao X, Zhou Q, Ma LQ (2006) Accumulation of Pb, Cu, and Zn in native plants growing on a contaminated Florida site. Sci Total Environ 368:456-464. doi:10.1016/j.scitotenv.2006.01.016

Zeb BS, Mahmood Q, Jadoon S, Pervez A, Irshad M, Muhammad B, Zulfiqar AB (2013) Combined industrial wastewater treatment in anaerobic bioreactor posttreated in constructed wetland. BioMed Res Int 2013:1-8.
http://dx.doi.org/10.1155/2013/957853

Zhu B, Panke-Buisse K, Kao-Kniffin J (2015) Nitrogen fertilization has minimal influence on rhizosphere effects of smooth crabgrass (*Digitaria ischaemum*) and bermuda grass (*Cynodon dactylon*). J Plant Ecol 8(4):390-400. doi:10.1093/jpe/rtu034

Ch. 8. Performance evaluation of duplex constructed wetlands for the treatment of diesel contaminated wastewater

This chapter has been submitted for publication as:

Hassana Ibrahim Mustapha., Pankaj Kumar Gupta., Brijesh Kumar Yadav., J.J.A van Bruggen., P.N.L Lens 2018. "Performance evaluation of duplex constructed wetlands for the treatment of diesel contaminated wastewater". Chemosphere, DOI:10.1016/j.chemosphere.2018.04.038. PII: S0045-6535(18)30678.7.

Abstract

A duplex constructed wetland (duplex-CW) is a hybrid system that combines a vertical flow (VF) CW as a first stage with a horizontal flow filter (HFF) as a second stage for a more efficient wastewater treatment as compared to traditional constructed wetlands. This study evaluated the potential of the hybrid CW system to treat influent wastewater containing diesel range organic compounds varying from $C_7 - C_{40}$ using a series of 12-week practical and numerical experiments under controlled conditions in a greenhouse (pH was kept at 7.0 ± 0.2, temperature between 20 and 23° C and light intensity between 85 and 100-μmol photons m^{-2} sec^{-1} for 16 h d^{-1}). The VF CWs were planted with *Phragmites australis* and were spiked with different concentrations of NH_4^+-N (10, 30 and 60 mg/L) and PO_4^{3-}-P (3, 6 and 12 mg/L) to analyse their effects on the degradation of the supplied petroleum hydrocarbons. The removal rate of the diesel range organics considering the different NH_4^+-N and PO_4^{3-}-P concentrations were simulated using Monod degradation kinetics. The simulated results compared well with the observed database. The results showed that the model can effectively be used to predict biochemical transformation and degradation of diesel range organic compounds along with nutrient amendment in duplex constructed wetlands.

Keywords: Duplex-CWs, hybrid systems, refinery diesel wastewater, total petroleum hydrocarbons, Numerical experiments

8.1 Introduction

Discharges of wastewater by petroleum industries (Semrany et al., 2012; Farhadian et al., 2008) into (sub) surface water resources are an important source of water pollution. Total petroleum hydrocarbons (TPH) are the main pollutants present in these industries (Al-Baldawi et al., 2014; Chavan, et al., 2008). TPH refer to a broad family of chemical compounds in water, soil or air that indicate the petroleum content (Pawlak et al., 2008) and can cause hazards for human health (Farhadian et al., 2008). Their main constituents are diesel, petrol, benzene, toluene, ethylbenzene, xylene (BTEX) and kerosene (Balachandran et al., 2012). The focus of this study was to treat synthetic wastewater with the characteristics of diesel contaminated effluent. Diesel ($C_6 - C_{26}$) is a complex fuel mixture comprising of hundreds of organic compounds (Al-Baldawi et al., 2014), it is composed of 65 - 85% saturated hydrocarbons, 5 – 30% aromatic hydrocarbons and 0 – 5% olefin fractions (Eze and Scholz, 2008; Liang et al., 2005), though the percentage may vary with manufacturer, mining location, refining process, sulphur content (Liang et al., 2005) and source of the crude petroleum (Agarwal et al., 2013).

Wastewater from petroleum refining and petrochemical industries contain BTEX compounds, polyaromatic hydrocarbons (PAHs), oil and grease, large amounts of suspended particulate matter, sulphides, ammonia and phenol (Mustapha et al., 2015; Tobiszewski et al., 2012; Wake, 2005). These discharges are one of the major environmental hazards to humans and animals (Ribeiro et al., 2013a; Seeger et al., 2011). For instance, benzene can cause leukemia at concentrations greater than 1 μg L^{-1} in drinking water (Chen et al., 2012; van Afferden et al., 2011). Similarly, other BTEX compunds are also toxic in nature and are generally considered carcinogenic to humans (van Afferden et al., 2011; Farhadian et al., 2008). Furthermore, oily wastewater can lead to the loss of biodiversity, destruction of breeding habitats of aquatic organisms and hazard to biota, including humans (Ribeiro et al., 2013b; Mustapha et al. 2011).

Constructed wetlands (CWs) are a promising alternative to conventional remediation systems (Wu et al., 2015) for the treatment of industrial wastewater (van Afferden et al., 2011) due to their efficiency (Yadav et al., 2009), cost-effectiveness (Al-Baldawi et al., 2014; Chen et al., 2012), low energy requirement (Aslam et al., 2007) and environmental friendliness (Mustapha et al., 2015; Mathur and Yadav, 2009; Yang et al., 2005). They are easy to operate and maintain (Al-Baldawi et al., 2014) and can simultaneously treat multiple contaminants

(Pardue et al., 2014;Yadav et al. 2011) from wastewater. Plants used in CWs improve the pollutant removal efficiency by enhancing the microbial diversity and their metabolic activity. Further, the plants play an important role in improving microbial activity by improved supply of oxygen and release of root exudates in the rhizospheric zone, which ultimately enhances the degradation rate of target pollutants (Al-Baldawi et al., 2013; Chen et al., 2012; Seeger et al., 2011).

The hybrid CW is a combination of two or more CWs connected in series that conglomerates the advantages of single CW systems to provide a better effluent water quality. A combination of vertical flow (VF) and horizontal flow (HF) CWs can optimize organic and nitrogen removal due to the presence of aerobic, anaerobic and anoxic phases (Saeed et al., 2012). A duplex-CW is a hybrid CW composed of a VFCW on top of a HF filter (HFF) with the additional advantage of a reduced space requirement (Zapater-Pereyra et al., 2015). Practical experiments are required to investigate the effectiveness of these CW for treating polluted water under varying environmental conditions. Likewise, simulation experiments are required to increase the understanding in dynamics and functioning of the complex CW systems that describe the transformation and degradation kinetics in a conceptual way (Yadav and Hassanizadeh, 2011). Numerical experiments can also help in evaluating and improving the existing design criteria of CW.

Numerical approaches are available to evaluate the performance of subsurface flow constructed wetlands, especialy HFF CWs (Al-Baldawi et al., 2014; Pastor et al., 2003). Pastor et al. (2003) used a hybrid neural network model to investigate the degradation of different pollutants in water through HFF CWs. Similarly, Tomenko et al. (2007) evaluated the effeciency of HFF CWs using artificial neural network models. Małoszewski et al. (2006) used tracer experiments to determine hydraulic parameters in three parallel gravel beds at a HFF CW in Poland. Wynn and Liehr (2001) predicted the seasonal trends in the removal efficiencies of HF CWs. The model consisted of six linked submodels representing the carbon and nitrogen cycles, an oxygen balance, autotrophic bacterial growth, heterotrophic bacterial growth, and a water budget. Further, Mayo and Bigambo (2005) developed a model to predict nitrogen transformations in HFF CWs.

Most of these studies either consider a single component or small domain processes in HFF CWs. Only a few modules like the constructed wetland 2D (CW2D) developed by Langergraber and Šimůnek (2006) simulated multi-component reactive transport of pollutants along with variably saturated water flow through the VF and HFF CWs. This module is incorporated with the HYDRUS 2D model (Šimůnek et al., 2006) to solve the soil moisture flow equation for variably saturated zones and the convection–dispersion equation for heat and contaminant transport simulations. The contaminant transport equation incorporates a sink term to account for the degradation or plant uptake. The CW2D has been successfully applied to model constructed wetlands for several pollutants including organic pollutants like NAPLs (Henrichs et al., 2007; Dittmer et al., 2005). Degradation by two types of bacteria (heterotrophic and autotrophic bacteria), which are responsible for the overall attenuation of organic pollutants, hydrolysis, and denitrification in CWs can be considered in CW2D (Langergraber and Šimůnek, 2006). Furthermore, CW2D is able to model the role of organic matter, nitrogen and phosphorus on the biochemical elimination and transformation processes of organic pollutants (Dittmer et al., 2005).

In this study, the performance of duplex-CWs having a VF and HFF domain treating a synthetic diesel wastewater was simulated using the CW2D to investigate the degradation of target pollutants. The objectives of this study are to (1) evaluate the performance of duplex-CWs in treating hydrocarbon $C_7 - C_{40}$ diesel range compounds in synthetic petroleum diesel contaminated wastewater under a continous flow and (2) investigate the performance of the three duplex-CWs spiked with different concentrations of nutrients on organic contaminant removal using a series of practical and numerical experiments.

8.2. Material and Methods

8.2.1 Experimental setup

The experimental setup consisted of three evenly spaced laboratory scale duplex-CWs, three influent tanks having a capacity of 200 L made of high-density polyethylene plastic containers located next to the duplex-CWs, and three peristaltic pumps used to pump the influent into the duplex-CWs. The duplex-CWs were composed of two parts: 1) a VF CW planted with common reed *Phragmites australis* and 2) an unplanted horizontal flow filter (HFF). The hybrid CWs were placed in a climate-controlled greenhouse (Zapater-Pereyra et al., 2015). The temperature of the greenhouse was kept between 20 and 23 ° C and a light intensity between 85 and 100 μmol photons m^{-2} sec^{-1} for 16 h d^{-1} was maintained throughout the experiments. During the later days of the experiment, days 35-56, there was a discontinuity in the lighting of the greenhouse chamber that affected the temperature of the treatment system.

The VF of the duplex-CWs had dimensions of 0.6 m L x 0.4 m W x 0.8 m D, while the HFF had dimensions of 0.6 m L x 0.4 m W x 0.35 m D with a surface area of 0.24 m² (L x W) for each set up. The VF was filled with 10 cm depth of gravel (15 – 30 mm) and was covered with 70 cm fine sand (1–2 mm) on top (Fig 8.1). The entire depth of the HFF was filled with 35 cm of fine sand as shown in Fig. 8.1. The porosity of the porous media used in the duplex-CW, hydraulic residence time (HRT), and effective volume of the treatment setups were estimated based on the dimensions of the CWs. The properties of the porous media estimated in primary studies are listed in Table 8.1. The VF and the HFF CWs had an effective volume of 0.0768 m³ and 0.0336 m³ and a hydraulic retention time of 4.8 days and 2.1 days, respectively. Before starting the data collection, the duplex-CWs were allowed to acclimatise for approximately 4 weeks (28 days). Thereafter, the data collection was conducted which was done for an additional 56 days.

All three duplex-CWs were operated under identical conditions. However, nutrient levels were modified after two weeks by adding different concentrations of mineral nitrogen (NH_4Cl) and phosphate (K_2HPO_4) to each of the influent tanks (INF1, INF2 and INF3), corresponding to concentrations of 10, 30 and 60 mg/L of NH_4^+-N and 3, 6 and 12 mg/L of PO_4^{3-}-P, respectively between days 21 to day 56.

Table 8.1. Characteristics of the practical and simulation experimental domain used to investigate the degradation of diesel compounds in the duplex-CWs.

Characteristics	Values	Source
Porosity (gravel)	40%	Determined
Porosity (sand)	34 %	
Effective volume	0.0768 m³	Calculated
Hydraulic retention time (HRT)	4.85 days	Calculated
Hydraulic loading rate (HLR)	0.0660 m³/m²/day	Calculated
Diffusivity (D^*)	6.3×10^{-6} [cm²/sec]	Calculated
Dispersion coefficient (D_i)	3.4 ± 0.2 [cm²/h]	
Flow rate	0.0158 m³/day	Calculated
Total simulation time	56 days	Experimental duration
Maximum growth rate μ_{max}	6 [1 d⁻¹]	Langergraber and Šimůnek, 2005
Growth limiting concentration K_s	0.5 [mgL⁻¹]	Langergraber and Šimůnek, 2005

8.2.2. Preparation of synthetic diesel refinery oil effluent

The synthetic diesel oil refinery wastewater (SDRW) contained 10 – 60 mg/L ammonium nitrogen, 0.3 – 12 mg/L orthophosphate and 0.4 mg/l phenol. The pH of the SDRW was kept around 7.0 (± 0.2). Thereafter, 40 mL of diesel was introduced directly into the tank and deionized water was added to bring SDRW to the desired level of 200 L. the tank was aerated to properly mix the influent. The tanks were washed thoroughly every 5 days, before fresh SDRW was prepared to prevent degradation by algae and microbes in the feeding tank.

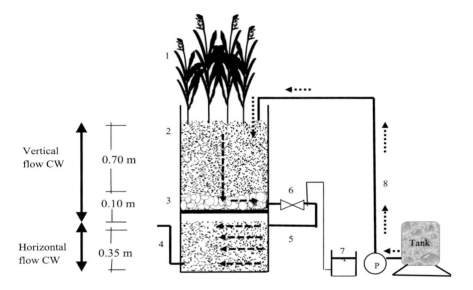

Fig. 8.1. Schematic representation of the duplex-CW configuration used in this study. 1-Phragmites australis, 2- Sand, 3- Gravel, 4- Outlet pipe, 5- Pipe connecting the compartments, 6- Valve, 7- Effluent collection bucket, 8- Inlet pipe and 9 - Influent tank.

8.2.3 Wastewater sampling and analysis

Direct measurements were taken for pH (pH meter; WTW pH 340), temperature and electrical conductivity (conductivity meter; WTW 82362) as well as dissolved oxygen (Wrinkler method) twice a week for the VF and HFF CWs. Samples were collected in 1 L calibrated bottles for laboratory analyses. Laboratory analyses were carried out twice a week using standard methods for the examination of water and wastewater (APHA, 2002): the closed reflux method was used for the determination of COD; UV-250 IPC (UV-VIS recording Spectrophotometer, Shimadzu) was used for the determination of NH_4^+-N and ICS – 1000, Ion Chromatography System, Dionox for NO_3^--N and PO_4^{3-}. GCMS was used to measure the concentration of the petroleum compounds (C_7 – C_{40}, benzene, toluene, ethylbenzene, m/p xylene and o-xylene) present in the synthetic diesel wastewater and the effluents from the duplex-CW treatment systems. The samples for Gas chromatography-mass spectrometry (GC-MS) were stored in a cold room at $5°C$ because the measurements of these parameters could not be done immediately.

8.2.4 Numerical Experiments

Numerical experiments were performed to investigate the biodegradation of diesel compounds in a synthetic contaminated wastewater in the duplex constructed wetlands using CW2D. A 2D study domain representing both VF and HFF of duplex-CW was created first to numerically solve the processes of the treatment system (Fig. 8.2). Fig. 8.2 shows the numerical domain of the duplex CW having the VF and HFF along with their material distribution. The study domain of the VF CW had dimensions of 0.8 m × 0.4 m, whereas that of the HFF had dimensions of 0.35 m × 0.4 m for D × W, respectively. The top layer of 0.7 m and undelaying 0.1 m of the VF domain was assigned with sand and gravel characteristics, respectively. Likewise, the entire HFF domain was assigned with sand medium only. The simulation domain was discretised in small fine element grids of hexahedral geometry, having a size of 1 cm with a stretching factor of unity for solving the governing equations numerically. Three observation nodes situated at the top and outlet of the VF and HFF were considered to represent the sampling ports of the experimental setup.

The hydraulic properties of the porous medium were characterized using the van Genuchten-Mualem approach (van Genuchten, 1980). The Crank-Nicholson iterative scheme was used for time weighting solution for every new time step of the nonlinear nature of the governing equations. The initial condition for the simulation domain was specified as saturated moisture content of the respective porous media. For the solute transport simulation, zero concentration of the selected pollutants was taken as an initial condition.

All three duplex CW were differentiated by their nutrient levels incorporated as initial concentrations of nitrogen and phosphate to each of the influent tanks (INF1, INF2 and INF3), corresponding to concentrations of 10, 30 and 60 mg/L of NH_4^+-N and 3, 6 and 12 mg/L of PO_4^{3-}-P, respectively. No flux boundary was taken at the lower and side faces of the tank set ups. The outlet point of the HFF was considered as a free drainage element. Surrounding greenhouse conditions were taken as the top atmospheric boundary condition of all VF.

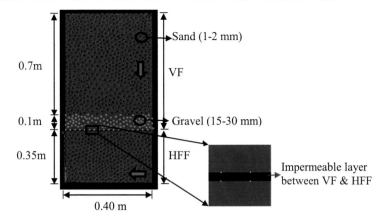

Fig. 8.2. Simulation domain of duplex-CW set up used in this study. The three red dots show the observation nodes at the top of the VF, the outlet of the VF, and outlet of the HFF to represent the sampling ports of the laboratory set up.

Autotrophic bacteria were assumed to be accountable for nitrification which was simulated as a two-step process, representing ammonium and nitrate oxidation. Other wetland processes like hydrolysis, aerobic growth and denitrification (anoxic growth) were simulated considering heterotrophic bacteria in the duplex-domain. The Monod type bio-kinetic model was used in mass balance equations to incorporate the function of the autotrophic and heterotrophic bacteria in all three duplex-CWs. The Monod kinetics can describe degradation rates of hydrocarbons including diesel range organics (DRO) for varying concentrations to represent zero to first order kinetics (Yadav et al. 2011) which can be written as:

$$\frac{\partial C}{\partial t} = k_m C / (K_s + C) \qquad (8.1)$$

where: rate constant (k_m) equals to $\mu_{max} X_0$, K_s the growth limiting concentration and C the target compound concentration at time t. The growth rate depends on the maximum growth rate of the autotrophic/heterotrophic bacteria (μ_{max}) and the concentration required to produce the initial biomass concentration (X_0). This kinetic expression was also used to simulate a) the sequential nitrification of ammonium into nitrite and nitrate by *Nitrobacter* and *Nitrosomonas*, respectively; b) nitrite and nitrate based growth of heterotrophs (denitrification) and c) hydrolysis (Langergraber and Šimůnek, 2005).

8.3 Results

8.3.1 Water quality parameters

The mean characteristics of the influent and effluents from the two stages of the duplex-CWs are presented in Table 8.2. A decrease in pH of the effluents of stage 1 and stage 2 was observed in all the CWs. DO concentrations were relatively lower in the effluents generated from stage 1 and 2. Also, the electrical conductivity of the treated effluent at stage 1 (VF1 – VF3) and stage 2 (HFF1 – HFF3) was higher than the electrical conductivity observed in the influents (INF1 – INF3).

8.3.2 Organic pollutants

8.3.2.1 Chemical oxygen demand

The mean COD influent and effluent concentrations are presented in Table 8.3 and Fig. 8.3. The COD results showed a decline in performance and then improved gradually from 63 to 84% for VF1, 57 to 85% for VF2 and 43 to 86% for VF3 when nutrients were added to the influents of the duplex-CWs (Fig. 8.3). The overall removal efficiencies from the combined CWs of the duplex – CWs ranged from 53.8 (± 33.1) to 69.4 (± 29.7)% (Table 8.3). However, duplex-CW1 had the best performance for COD removal and duplex CW3 had the poorest mean COD removal efficiency. The maximum COD removal efficiency in stage 1 was by VF1 (97%) and in stage 2 by HF1 (98%), despite VF1 and HF1 were supplied with the lower nutrient concentrations (10 mg NH_4^+-N/L and 3 mg PO_4^{3-}-P/L).

Table 8.2. *Mean composition of the influent at different nutrient concentrations and effluent samples from the 2-staged duplex CWs with minimum and maximum values (n = 32).*

	*Mean ± SD	Minimum	Maximum
Duplex CW 1 (8 mg/L of nitrogen and 3 mg/L of phosphorus)			
Influent (INF1) – System 1			
pH	7. 7 ± 0.2	7.4	7.9
Temperature (°C)	16.4 ± 4.1	11.9	21.1
Dissolved oxygen (mg/L)	9.7 ± 0.5	8.8	10.1
Electrical conductivity (μS/cm)	596.5 ± 64.8	512.5	650.0
Influent – stage 1 (VF CW1)			
pH	6.6 ± 0.3	6.2	7.0
Temperature (°C)	16.8 ± 3.3	9.9	19.8
Dissolved oxygen (mg/L)	4.6 ± 1.4	2.1	6.9
Electrical conductivity (μS/cm)	617.9 ± 73.7	542.5	747.0

Effluent - Stage 2 (HFF1)

pH	6.9 ± 0.2	6.6	7.2
Temperature (°C)	16.2 ± 3.1	11.8	19.5
Dissolved oxygen (mg/L)	3.0 ± 0.7	1.9	3.9
Electrical conductivity (µS/cm)	646.8 ± 72.0	567.0	756.0

Duplex CW 2 (31 mg/L of nitrogen and 6 mg/L of phosphorus)
Influent (INF2) – System 2

pH	7.7 ± 0.2	7.5	8.0
Temperature (°C)	16.3 ± 4.1	11.6	20.9
Dissolved oxygen (mg/L)	9.7 ± 0.5	8.6	10.1
Electrical conductivity (µS/cm)	713.1 ±162.8	517.0	856.5

Influent – stage 1 (VF CW2)

pH	6.5 ± 0.4	6.1	7.0
Temperature (°C)	16.9 ± 3.4	9.8	19.6
Dissolved oxygen (mg/L)	4.6 ± 1.3	2.6	6.7
Electrical conductivity (µS/cm)	713.4 ± 129.8	556.0	848.0

Effluent - Stage 2 (HFF2)

pH	6.6 ± 0.2	6.2	6.9
Temperature (°C)	16.2 ± 3.1	11.8	19.5
Dissolved oxygen (mg/L)	3.2 ± 0.8	2.4	4.7
Electrical conductivity (µS/cm)	687.6 ± 129.8	526.0	793.0

Duplex CW 3 (66 mg/L of nitrogen and 15 mg/L of phosphorus)
Influent (INF3) – System 3

pH	7.7 ± 0.2	7.5	8.1
Temperature (°C)	16.1 ± 4.4	10.9	20.9
Dissolved oxygen (mg/L)	9.7 ± 0.5	8.6	10.0
Electrical conductivity (µS/cm)	917.4 ± 329.1	517.0	1187.0

Influent – stage 1 (VF3)

pH	6.5 ± 0.5	6.0	7.1
Temperature (°C)	16.8 ± 3.4	9.8	19.9
Dissolved oxygen (mg/L)	4.0 ± 1.3	2.8	6.3
Electrical conductivity (µS/cm)	881.9 ± 272.4	551.5	1134.5

Effluent - Stage 2 (HFF3)

pH	6.7 ± 0.2	6.5	6.9
Temperature (°C)	16.0 ± 3.1	11.7	19.5
Dissolved oxygen (mg/L)	2.8 ± 0.6	2.0	3.8
Electrical conductivity (µS/cm)	866.6 ± 253.4	559.0	1099.0

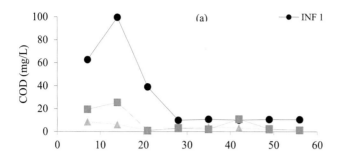

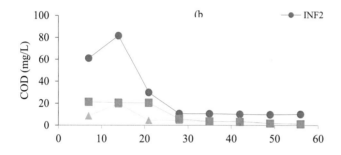

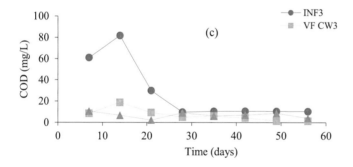

Fig. 8.3. Influent COD and effluents from the 2 stages (VF and HFF) during the experimental period from (a) duplex-CW 1, (b) duplex-CW 2 and (c) duplex-CW 3.

8.3.2.2 Diesel range organics

The influent and effluent diesel range hydrocarbon (DRO) from $C_7 - C_{40}$ for INF1, INF2 and INF3 are presented in Table 8.3. The corresponding observed removal efficiencies are presented in Table 8.4. The quantities of $C_7 - C_{40}$ alkane compounds were greatly decreased (>90%) in the effluents from the three duplex-CWs, irrespective of the quantity of nutrients applied. However, the stage 1 (VF component) of the duplex-CW3 had the highest mean performance (99.9%) followed by VF2 (98.3%) and then VF1 (97.8%). Meanwhile, duplex-CW2 had the highest mean performance (99.81%), followed by duplex-CW1 and then the duplex-CW3 for the hybrid system performance. Similarly, the simulated removal efficiencies varied from 84.4 ($\pm$ 2.4) to 95.6 ($\pm$ 0.1)% and from 80.8 ($\pm$1.3) to 94.7 ($\pm$ 0.5)% for stage 1 and stage 2, respectively. For the hybrid CWs, duplex CW3 had the lowest removal efficiency of 96.9 ($\pm$0.5)% and duplex CW2 had the highest DRO removal efficiency of 99.8 ($\pm$ 0.2)%. The simulated DRO removal efficiencies were 2.2, 0.3 and 3.0% lower than the observed values for duplex-CW1, duplex-CW2, and duplex-CW3, respectively (Table 8.2).

8.3.2.3. Benzene

The average benzene concentrations in the influent (INF1 – INF3), stage 1 (VF CW1 – VF CW3) and stage 2 (HFF1 – HFF3) are presented in Table 8.3. The mean removal efficiencies varied between 86 ($\pm$24.3) and 93 ($\pm$11.5) % for stage 1 (VF CW1 – VF CW3). The VF CW1 showed a 100% removal of benzene on the 49[th] and 56[th] day of the treatment run, while VF CW2 showed a 100% benzene removal efficiency on the 56[th] day and VF CW3 showed a 100% benzene removal at the 42[th] and 56[th] day of the treatment. There was no additional removal of benzene in the HFF (1 - 3) compartments (Table 8.3). These gave a corresponding removal efficiency of 77.9 ($\pm$ 3.6) %, 62.2 ($\pm$ 46.8) % and 53.3 ($\pm$ 46.2) % in the laboratory experiment and 98.4 ($\pm$ 0.6) %, 96.2 ($\pm$ 0.5) % and 97.4 ($\pm$ 1.7) % in the numerical experiments for duplex-CW1, duplex-CW2 and duplex-CW3, respectively (Table 8.4). The duplex-CWs treatment systems showed a good benzene degradation in the VF compartment. Also, the observed and simulated data indicated an almost 100% removal of benzene at day 56 by the three duplex-CWs at average temperatures above 16 °C and nutrient concentrations above 8 mg/L NH_4^+-N and 3 mg/L PO_4^{3-}-P.

8.3.2.4. Toluene

The influent toluene (INF1 – INF3) and mean effluent toluene concentrations for stage 1 (VF CW1 – VF CW3) and stage 2 (HFF1 – HFF3) are presented in Table 8.2. The removal efficiencies ranged from 88 to 92% for stage 1. The three VF CWs showed a 100% toluene removal at the 56th day of the treatment, while in the HFF (stage 2) compartments no additional removal was observed by the three duplex-CWs (Table 8.4). The mean toluene removal efficiencies from the laboratory setup and simulation domain were 58.3 ($\pm$ 52.0)%, 89.4 ($\pm$ 12.9)%, 88.3 ($\pm$ 12.6)% and 93.3 ($\pm$ 11.5)%, 100%, 100% for duplex-CW1, duplex-

CW2 and duplex-CW3, respectively. The duplex CW2 and CW3 showed the highest performance with 88.3 - 89.4% and 100% toluene removal from the experimental and simulation domains, respectively.

8.3.2.5. Ethylbenzene

The influent ethylbenzene (INF1 – INF3) and the effluent concentrations for stage 1 (VF CW1 – VF CW3) and stage 2 (HFF1 – HFF3) are presented in Table 8.3. The removal efficiencies ranged from 64.9 (± 56.3)% to 100 (± 0.0)% for the first stage (VF CWs). The VF CW1 had a 100% ethylbenzene removal efficiency from day 42 to day 56, while VF CW2 and VF CW3 had 100% removal efficiencies at day 56. The HFF (stage 2) compartments of the three duplex-CWs showed no additional ethylbenzene removal from day 42 to day 56 for the HFF1, while HFF2 and HFF3 showed no additional removal at day 56, which also corresponded to a 100% ethylbenzene removal efficiency by the VF CW 2 and VF CW3 at day 56. The observed and simulated removal efficiencies from the duplex-CWs were 100 (± 0.0)%, 91.7 (± 14.4)%, 88.9 (± 19.2)% and 100%, 91.7 (± 14.4)%, 86.7 (± 23.1)% for duplex-CW1, duplex-CW2 and duplex-CW3, respectively (Table 8.3). The observed and simulated removal efficiencies matched well for the duplex-CW1 and duplex-CW2. The highest ethylbenzene removal efficiency occurred in duplex-CW1, followed by duplex CW2 and duplex CW3.

8.3.2.6. m/p xylene

The influent m/p xylene (INF1 – INF3) and the effluent concentrations for stage 1 (VFCW1 – VFCW3) and stage 2 (HFF1 – HFF3) are presented in Table 8.3. All the three VF CWs reached a 100% m/p xylene removal efficiency at day 56. Meanwhile, the corresponding mean m/p xylene removal efficiencies ranged from 66.0 (± 57.1)% to 94.2 (± 6.3)% for the stage 1. Similarly, no additional m/p xylene removal was observed in the HFF1 and HFF3 compartments of the three duplex-CWs (Table 8.3) and from day 49 to day 56 for HFF2. Duplex-CW1, duplex-CW2 and duplex-CW3 had mean m/p xylene removal efficiencies of 88.0 (±12.5)%, 93.8 (± 9.1)% and 90.1(± 13.3)%, respectively. In the simulation domain, the removal efficiencies of 96.7 (± 5.8)%, 100% and 100% were obtained for duplex-CW1, duplex-CW2 and duplex-CW3, respectively.

8.3.2.7. o-xylene

The influent o-xylene (INF1 – INF3 and effluent concentrations from the stage 1 (VF CW1 – VF CW3) and for stage 2 (HF1 – HFF3) are presented in Table 8.3. The observed mean o-xylene removal efficiencies ranged from 66.7 (± 57.7) to 100 (± 0.0)% for stage 1 and from no additional removal to 32.8 (± 56.9)% for stage 2. Meanwhile, the VF CW1 and VF CW3 had a 100% o-xylene removal efficiency from day 42 to day 56, while a removal of 100% was observed in VF CW2 from day 49 to day 56. The duplex-CWs had a range of mean o-xylene removal efficiencies from 91.7 (± 14.4) % to complete o-xylene removal (100 (± 0.0)%). Likewise, simulated removal efficiencies varied from 60 - 93.3 (±11.5)% for VF CW. The no additional removal efficiencies in the HFF CW shown in the simulated results is due to lack of o-xylene present in the HFF influent after the 56th day. Furthermore, 100 % and 98.7% removal efficiencies are shown in duplex CW1 and CW2, respectively. Almost a 100% removal efficiency was found for o-xylene in duplex CW1 for the laboratory setup as well as for the simulation study domain.

8.3.2.8 Nutrients

The influent and effluent NH_4^+- N and NO_3^-- N concentrations are presented in Fig. 8.4 and Fig. 8.5. The mean NH_4^+- N removal efficiencies by the VF CWs (stage 1) ranged between 54.4 (± 25.6)% and 96.8 (± 4.1)% and between 6.2 (± 17.5)% and 28.8 (± 44.2)% by the HFF (stage 2) compartments (Table 8.4). The mean NO_3^-- N removal efficiencies by the VF CWs (stage 1) and by the HFF (stage 2) compartments ranged between 25.8 (± 33.1)% and 81.7 (± 33.3)% and 12.5 (± 34.3)% and 28.6 (± 41.0)%, respectively (Table 4). NH_4^+- N and NO_3^-- N removal were high in the duplex-CW1. The duplex-CWs had NH_4^+- N removal efficiencies of 80.9 (± 33.4)%, 82.0 (± 26.5)% and 66.8 (± 12.7)%, respectively, for duplex-CW1, duplex-CW2 and duplex-CW3. The NO_3^--N removal efficiencies from the duplex-CWs were 80.1 (± 33.9)%, 33.8 (± 46.2)% and 53.0 (± 45.3)% for duplex-CW1, duplex-CW2 and duplex-CW3, respectively.

Fig. 8.6 present the influent (INF1 – INF3) PO_4^{3-}-P and effluent concentrations for stage 1 and stage 2. The mean removal efficiencies varied from 40.0 (± 31.1)% to 49.8 (± 24.5)% for stage 1 and from 1.0 (± 2.7)% to 26.9 (± 45.4)% for stage 2 (Table 8.3). The mean PO_4^{3-}-P removal efficiencies from the duplex-CWs varied from 31.4 (± 30.2)% to 60.8 (± 45.4)%. The duplex-CW1 had the highest PO_4^{3-}-P removal efficiency followed by duplex-CW3, whereas the duplex-CW2 had the lowest PO_4^{3-} -P removal efficiency (Table 8.3).

Table 8.3. Composition of influent and effluent samples (±S.D, mg/L) from the components of the three two-stage duplex-CWs with minimum and maximum values

Parameter		INF1	VF CW1	HFF1	INF2	VF CW2	HFF2	INF3	VF CW3	HFF3
COD (mg/L)	Average	31.7±33.6	3.6±2.5	8.2±9.5	28.1±28.1	6.1±5.9	9.8±9.2	28.1±28.1	6.9±5.6	6.8±2.7
	Min	9.9	1.2	0.9	9.9	1.6	1.1	9.8	1.5	2.32
	Max	99.7	8.5	25.6	81.7	19.7	21.4	81.7	19.0	10.6
DRO (mg/L)	Average	84.3±1.2	1.8±2.4	0.2±0.0	246.3±198.7	2.1.±1.9	0.3±0.2	165.3±86.0	0.2±0.1	0.45±0.4
	Min	83	0.3	0.2	90	0.2	0.1	80	0.1	0.1
	Max	85	4.6	0.2	470	4.0	0.4	252	0.3	0.8
Benzene (mg/L)	Average	0.1± 0.1	0.0±0.0	0.0±0.0	0.2±0.2	0.1±0.0	0.1±0.1	0.1±0.0	0.0±0.0	0.0±0.1
	Min	0.1	0.0	0.0	0.1	0.0	0.0	0.1	0.0	0.0
	Max	0.2	0.1	0.1	0.4	0.0	0.2	0.2	0.0	0.1
Toluene (mg/L)	Average	0.03±0.02	0.00±0.01	0.01±0.01	0.07±0.07	0.01±0.01	0.01±0.01	0.06±0.03	0.01±0.01	0.01±0.01
	Min	0.01	0.00	0.00	0.03	0.0	0.0	0.04	0.0	0.0
	Max	0.04	0.01	0.01	0.15	0.01	0.01	0.10	0.01	0.01
Ethylbenzene (mg/L)	Average	0.02±0.01	0.00±0.0	0.00±0.00	0.09±0.09	0.02±0.03	0.0±0.01	0.05±0.03	0.0±0.01	0.0±0.01
	Min	0.02	0.00	0.00	0.03	0.0	0.0	0.03	0.0	0.0
	Max	0.03	0.00	0.00	0.19	0.05	0.01	0.08	0.01	0.01
m/p-xylene (mg/L)	Average	0.06±0.03	0.01±0.01	0.01±0.01	0.22±0.22	0.07±0.11	0.01±0.01	0.12±0.08	0.01±0.01	0.0±0.01
	Min	0.04	0.00	0.00	0.07	0.0	0.0	0.08	0.0	0.0
	Max	0.09	0.01	0.01	0.48	0.19	0.02	0.21	0.01	0.02
o-xylene (mg/L)	Average	0.03±0.01	0.00±0.0	0.0±0.0	0.10±0.11	0.22±0.38	0.0±0.01	0.06±0.03	0.0±0.0	0.0±0.01
	Min	0.02	0.0	0.00	0.03	0.0	0.0	0.04	0.0	0.0
	Max	0.04	0.0	0.0	0.23	0.66	0.01	0.10	0.0	0.01

Note: INF - Influent; VF CW – vertical flow constructed wetland; HFF - horizontal flow filter; 1, 2 and 3 – set up 1, 2 and 3.

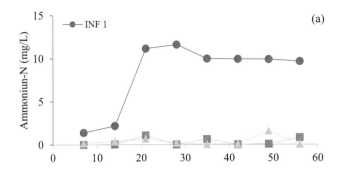

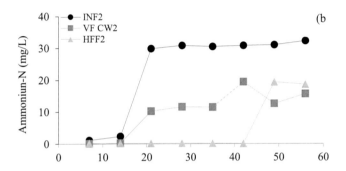

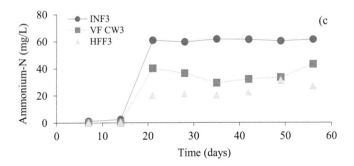

Fig. 8.4. Influent and effluent NH₄⁺-N concentrations from the 2 stages (VF and HFF) during the experimental period from (a) duplex-CW 1, (b) duplex-CW 2 and (c) duplex-CW 3.

Table 8.4. Performance of the 2-stage duplex CWs without nutrient supplement (Mean ± SD) (%) during the first 20 days of operation

Parameter	VF CW1	HFF1	Duplex (hybrid) CW1	VF CW2	HFF2	Duplex (hybrid) CW 2	VF CW3	HFF3	Duplex (hybrid) CW 3
COD	92.4±5.5	8.5±14.8	80.4±15.3	82.0±5.3	0±0	57.1±22.8	77.3±8.7	46.5±40.6	88.9±5.4
DRO	78.6±0.2	64.2±1.4	80.2±0.4	84.0±0.2	75.8±0.5	85.4±0.6	84.0±0.2	75.8±0.5	85.4±0.6
Benzene	67.5±0.2	23.8±2.4	65.0±0.2	63.1±0.4	22.8±0.2	62.0±0.2	62.4±0.5	22.8±0.2	60.0±0.5
Toluene	63.1±0.5	32.2±0.2	72.0±0.2	75.2±0.5	34.2±0.2	85.0±0.2	75.2±0.5	34.2±0.2	85.0±0.2
Ethylbenzene	89.0±0.5	14.2±0.2	90.5±0.2	85.2±0.5	08.2±0.5	87.2±0.5	85.2±0.5	08.2±0.5	87.2±0.5
m/p-xylene	48.0±0.5	78.0±0.5	81.5±0.5	52.5±0.2	72.5±0.5	83.5±0.5	50.5±0.2	78.5±0.5	85.5±0.5
o-xylene	58.5±0.2	63.2±0.5	74.5±0.5	64.5±0.2	61.2±0.5	78.5±0.5	64.5±0.2	61.2±0.5	78.5±0.5

Table 8.5. Performance of the 2-stage duplex CWs with nutrient supplement (Mean ± SD) (%) from day 21 to 56

Parameter	VF CW1	HFF1	Duplex (hybrid) CW1	VF CW2	HFF2	Duplex (hybrid) CW 2	VF CW3	HFF3	Duplex (hybrid) CW 3
COD	74.4±9.1	13.3±20.6	62.4±17.2	64.2±20.3	0±0	32.7±20.6	69.1±17.4	17.4±32.4	53.8±33.1
DRO*	97.8±2.8	65.2±31.2	99.8±0.0	98.3±2.4	75.5±22.1	99.8±0.2	99.9±0.1	NA	96.6±0.5
DRO#	86.4±2.4	81.9±1.3	97.6±0.3	95.5±0.1	90.0±0.5	99.5±0.0	91.5±0.4	94.3±0.5	99.5±0.1
Benzene*	86.0±24.3	NA	77.9±3.6	90.3±10	NA	62.2±46.8	93.3±11.5	NA	53.3±46.2
Benzene#	90.4±4.5	81.9±6.4	98.4±0.6	78.3±2.8	82.5±0.1	97.4±0.5	86.3±3.0	81.9±7.9	96.2±1.7
Toluene*	91.7±14.4	NA	58.3±52.0	89.4±12.9	NA	89.4±12.9	88.3±12.6	NA	88.3±12.6
Toluene#	53.3±5.8	83.3±28.9	93.3±11.5	79.3±20.0	NA	100±0.0	83.8±14.7	NA	100±0.0
Ethylbenzene*	100±0.0	NA	100±0.0	64.9±56.3	NA	91.7±14.4	95.8±7.2	NA	88.9±19.2
Ethylbenzene#	80.6±17.3	NA	100±0.00	55.6±9.6	NA	91.7±14.4	NA	NA	86.7±23.1
m/p-xylene*	88.0±12.5	NA	88.0±12.5	66.0±57.1	NA	93.8±9.1	94.2±6.3	NA	90.1±13.3
m/p-xylene#	50.0±0.0	93.3±11.5	96.7±5.8	33.3±0.0	NA	100.0±0.0	92.5±6.6	NA	100±0.0
o-xylene*	100±0.0	NA	100±0.0	66.7±57.7	NA	93.3±11.5	100±0.0	NA	91.7±14.4
o-xylene#	93.3±11.5	NA	100±0.0	86.7±23.1	NA	98.7±2.3	NA	NA	NA

Table 8.6. Removal efficiencies (Mean ± SD) (%) by the 2-stage duplex CWs from day 0 to 56

Parameter	VF CW1	HFF1	Duplex (hybrid) CW1	VF CW2	HFF2	Duplex (hybrid) CW2	VF CW3	HFF3	Duplex (hybrid) CW3
COD	81.1±12.0	11.5±17.7	69.4±29.7	74.8±11.8	6.2±11.7	64.8±18.9	69.1±17.4	17.4±32.4	53.8±33.1
C7-C40*	97.8±2.8	65.2±31.2	99.8±0.0	98.3±2.4	75.5±22.1	99.8±0.2	99.9±0.1	NA	96.6±0.5
C7-C40#	86.4±2.4	81.9±1.3	97.6±0.3	95.5±0.1	90.0±0.5	99.5±0.0	91.5±0.4	94.3±0.5	99.5±0.1
Benzene*	86.0±24.3	NA	77.9±3.6	90.3±10	NA	62.2±46.8	93.3±11.5	NA	53.3±46.2
Benzene#	90.4± 4.5	81.9±6.4	98.4±0.6	78.3±2.8	82.5±0.1	97.4±0.5	86.3±3.0	81.9±7.9	96.2±1.7
Toluene*	91.7±14.4	NA	58.3±52.0	89.4±12.9	NA	89.4±12.9	88.3±12.6	NA	88.3±12.6
Toluene#	53.3±5.8	83.3±28.9	93.3±11.5	79.3±20.0	NA	100±0.0	83.8±14.7	NA	100±0.0
Ethylbenzene*	100±0.0	NA	100±0.0	64.9±56.3	NA	91.7±14.4	95.8±7.2	NA	88.9±19.2
Ethylbenzene#	80.6±17.3	NA	100±0.00	55.6±9.6	NA	91.7±14.4	NA	NA	86.7±23.1
m/p-xylene*	88.0±12.5	NA	88.0±12.5	66.0±57.1	NA	93.8±9.1	94.2±6.3	NA	90.1±13.3
m/p-xylene#	50.0±0.0	93.3±11.5	96.7±5.8	33.3±0.0	NA	100.0±0	92.5±6.6	NA	100±0.0
o-xylene*	100±0.0	NA	100±0.0	66.7±57.7	NA	93.3±11.5	100±0.0	NA	91.7±14.4
o-xylene#	93.3±11.5	NA	100±0.0	86.7±23.1	NA	98.7±2.3	NA	NA	NA
Ammonium-N	96.8±4.1	28.8±44.2	80.9±33.4	64.8±18.1	12.5±35.2	82.0±26.5	54.4±25.6	6.2±17.5	66.8±12.7
Nitrate-N	81.7±33.3	25.4±39.2	80.1±33.9	25.8±33.1	12.5±34.3	33.8±46.2	47.2±42.8	28.6±41.0	53.0±45.3
Phosphate-P	40.0±31.1	26.9±45.4	60.8±45.4	42.3±28.5	1.0±2.7	31.4±30.2	49.8±24.5	12.4±35.2	56.0±23.2

NA- Not Applicable * Observed   # Simulated

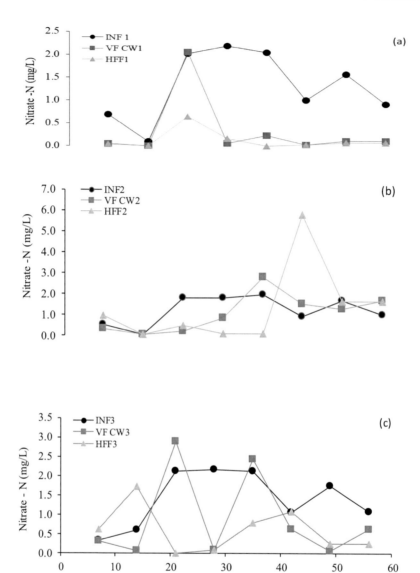

Fig. 8.5. Influent and effluents NO_3^--N concentrations from the 2 stages (VF and HFF) of the duplex CW during the experimental period from (a) duplex-CW1, (b) duplex-CW2 and (c) duplex-CW3.

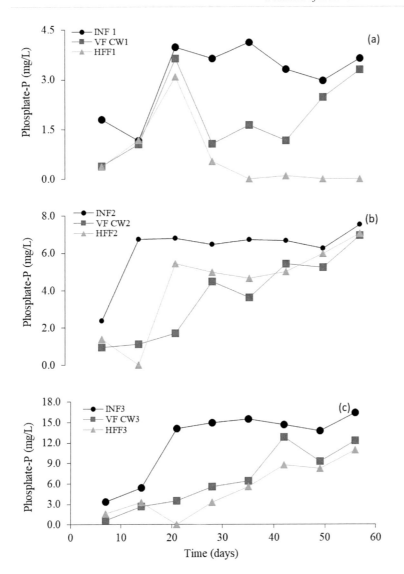

Fig. 8.6. Influent and effluents PO$_4^{3-}$ -P concentrations from the 2 stages (VF and HFF) of the duplex CW during the experimental period from (a) duplex-CW 1, (b) duplex-CW 2 and (c) duplex-CW 3.

8.4 Discussion

8.4.1. Constructed wetlands for treatment of diesel contaminated wastewater

This study showed that hybrid duplex-CWs can be used for the removal of diesel range organic compounds from simulated diesel wastewaters. The hybrid CW wetlands effectively reduced the concentrations of the contaminants to safe discharge limits for discharge into waterbodies or the environment (Tables 8.3, 8.4, 8.5, 8.6 and Fig. 8.3, 8.4 and 8.5).

DRO

A high removal efficiency (i.e. >95%) was observed for DRO in the simulated duplex-CWs domain treating diesel contaminated wastewater. The simulated removal efficiencies of duplex-CW1, duplex-CW2 and duplex-CW3 were comparatively lower than the respective observed values. This could be due to ignoring sorption of DRO on gravel material and side walls of the wetland setups. Similarly, Al-Isawi et al. (2015) investigated removal of diesel by VF CWs using a numerical approach and showed effective performance in attenuating these compounds from wastewater. The removal of DRO compounds from duplex-CWs was uniform throughout the simulation domain, which indicates that the duplex-CWs were able to maintain optimal oxidative/reductive conditions for nitrogen species (nitrification/denitrification) in the VF/HFF domain (Seeger et al., 2011).

Benzene

The consideration of atmospheric boundary conditions in the simulation domain causes continuous supply of oxygen in CWs, especially in VF CWs (Yadav and Hassanizadeh, 2011). The high porosity in the gravel zone of the VF was also equally contributing in maintaining the oxygen concentration. Thus, in this study all three VF CWs were able to maintain the optimal oxygen concentration in the domain and resulted in high benzene removal efficiencies by the VF CWs. The high performance of VF CWs caused a very low benzene concentration of entering the HFF CWs and ultimately resulted in the low removal efficiency of benzene in HFF CWs due to substrate limitation (Gupta and Yadav, 2017). This indicates that associated heterotrophic bacteria and/or acetotrophic methanogenic archaea in the CW2D generated numerical HFF CWs domain were not able to attenuate effectively at very low concentrations, even after nutrient supplementation (Langergraber and Šimůnek, 2005). Also ammonium oxidation can play a crucial role in benzene removal by stimulating ammonium oxidizing bacteria (autotrophic bacteria) in VF CWs.

Toluene

The simulation domain representing duplex-CW2 and CW3 shows complete degradation of toluene due to the comparatively high nutrient dose as compared to CW1. This high removal efficiency of toluene is apparently due to biostimulation or aerobic growth of heterotrophic bacteria as fortified by the findings of Yadav et al. (2014). Similarly, Eckert and Appelo

(2002) observed that the applied nitrate concentration accelerated the bio-kinetic reactions in the duplex-CWs. Thus, the VF-CWs simulation domain shows a high performance of toluene removal verifying that the considered processes are suitable to represent the wetland setup accurately. Langergraber and Šimůnek (2005) also considered Monod kinetics to predict the biochemical transformation and degradation of these compounds along with nutrient amendment in subsurface flow CWs successfully. The HFF CWs show a lower removal efficiency than VF, this may be due to the lower oxygen levels in the HFF CWs domain than in the VF CWs domain (Yadav et al., 2014).

Ethylbenzene

The simulated and observed removal efficiencies for ethylbenzene matched well for all three duplex-CWs. The similarity in ethylbenzene removal efficiencies by all three duplex-CW confirm that both the laboratory and simulation domain are well designed (Toscano et al., 2009). The highest performance of the duplex-CW1 followed by duplex-CW2 shows that the additional nutrients provided do not affect the degradation of ethylbenzene significantly. This may be due to the low concentration of ethylbenzene in the domain because BTEX compounds degradation kinetics depend on the initial substrate concentrations (Gupta et al., 2013).

Xylene

The removal of m/p-xylene in the VF-CWs was faster than the other BTEX compounds as compared to the HFF CWs. During the anaerobic electron accepting conditions, denitrifying conditions (i.e., where nitrate is the primary electron acceptor) were clearly the most helpful for the anaerobic m/p-xylene degradation and is in agreement with findings from Burland and Edwards (1999). The increased oxygen supply from the top boundary and application of nutrient dose to the VF CWs seems helpful in accelerating the biodegradation of m/p/o-xylene compounds as found by Basu et al. (2015).

The process based modelling approach applied in this study can thus be used to forecast the outflow concentration and possible changes in treatment performance when the influent concentration load is expected to vary in the influent of duplex-CWs (Rizzo and Langergraber, 2016; Meyer et al., 2015). Furthermore, consideration of substrate concentration and its species based biodegradation kinetic models will help to simulate more realistic conditions in similar treatment studies. However, other aspects like heterogeneity of the root zone and variations in atmospheric boundary conditions may be incorporated in the modelling framework in order to further predict the fate and transport of such types of pollutants more accurately.

8.4.2. Single stage versus hybrid CW system

This study showed that hybrid-CWs can achieve higher removal efficiencies than single stage VF CW while treating simulated diesel contaminated wastewater. The duplex-CWs showed higher removal efficiencies of DRO, toluene, ethylbenzene, m/p xylene and o-xylene than the VF CWs with more distinct observations in the no nutrient supplemented period (Table 8.5). The duplex-CWs had COD removal efficiencies of 80, 57 and 89%, respectively for duplex-CW1, 2 and 3 for the no nutrient supplemented treatment, these values are within the range obtained by Herrera Melián et al. (2010). The latter study reported a 83% COD removal for a gravel substrate with a hydraulic retention time of 4.1 h for the VF and 12 h for the HF CWs in the warm climate of the Canary Islands (Spain). The duplex-CW3 showed a higher COD removal efficiency as compared to the single systems (VF CW3/ HFF3) (Table 8.4) due to the combined effects of the VF CW3 - HFF CW3. Firstly, the VF CW provided sufficient amounts of dissolved oxygen through the incoming wastewater load (Table 8.2) for the mineralization of organic matter (Mustapha et al., 2015). The VF CWs also received the full organic load, since a higher organic load increases the performance of a CW (Herrera Melián et al., 2010), while the HFF CWs may further contribute to the anaerobic decomposition of the organic matter (Vymazal, 2005).

The three duplex-CWs had a range of 50 - 100% removal efficiency for all the contaminants investigated, i.e. benzene, toluene, ethylbenzene, m/p xylene and o-xylene. This removal efficiency may be due to simultaneous occurrence of biodegradation, sorption, hydrolysis and photodegradation processes within the hybrid duplex-CWs (Ávila et al., 2014). In this study, the main part of the overall removal of all the DRO contaminants occurred in the VF CWs. This may be related to the unsaturated conditions of the VF CWs, which correspond to oxidizing conditions, thus favouring aerobic microbial processes (Ávila et al., 2014; Hijosa-Valsero et al., 2011).

Ammonium concentrations were reduced in all the three duplex-CWs. The VF CWs played an important role in the reduction of ammonium concentrations (>50%) in the simulated refinery wastewater (Table 8.3). The aerobic conditions present in the VF CWs may have contributed to the ammonium removal as a result of the oxidation of NH_4^+-N to NO_3^- (Zurita and White, 2014). According to Ye et al. (2009), a DO above 1.5 mg/L is required for nitrification to occur. The three duplex-CWs had a DO above 1.5 mg/L (Table 8.2) in their effluent, thus nitrification may be the main mechanism responsible for the ammonium removal in the VF compartments. The ammonium removal from the three HFFs compartments was very low (6 - 29%) compared to results observed in single and hybrid CWs of other studies (Mustapha et al., 2015, Ye at el., 2009, Masi and Martinuzzi, 2007). This may suggest that the nitrification/denitrification processes were limited by inadequate microbial activity in the unplanted HFFs mineral medium (Drizo et al., 2000) as well as the absence of plants.

There were high variabilities in the nitrate concentration of the effluents of the three duplex-CWs. Low concentrations of nitrate in the influent as well as the effluents of the three duplex-CWs may be related to the limited nitrification capacity of the HFF compartments and as a result of effective nitrification-denitrification of NH_4^+-N to gaseous nitrogen (Seeger et al., 2011). The duplex-CW1 had a lower nitrate concentrations in both the VF CW1 and HFF1 as compared to duplex-CW2 and CW3 (Fig. 8.5). The nitrate produced in the VF CW1 was successfully reduced in HFF1. According to Kyambadde et al. (2005), lower effluent nitrate concentrations in planted CWs are attributed to a higher competition for oxygen between aerobic heterotrophic and autotrophic nitrifying bacteria, plant uptake of nitrate and denitrification. The higher nitrate concentrations in the effluents of duplex-CW2 and CW3 (Fig. 8.5) is an indicator of active nitrification (Kyambadde et al., 2005) in some days and lower nitrate concentrations in the effluents in the remaining days suggest a limited nitrification capacity (Seeger et al., 2011). Seeger et al. (2011) confirmed that microbial transformation (nitrification/denitrification) and plant uptake of ammonium or nitrate are the main N removal processes in CWs. However, 5 mg/L of NO_3^--N was recorded in stage 2 from the effluent of duplex-CW2 after day 42, which could be attributed to increased oxygen solubility at lower temperatures (Seeger et al., 2011). HF CW3 had greater reduction rates for nitrate compared to its ammonium removal rate. This is probably due to its lower level of dissolved oxygen (Zurita et al., 2014). Thus, denitrification may be responsible for the NO_3^--N removal at this HFF stage.

This study showed that duplex-CWs (1 and 3) had a higher mean phosphate removal efficiency than the single CWs (VF and HFF), confirming the advantages of using hybrid CW systems to provide a better effluent. Also, the VF CWs had higher removal efficiencies than the HFF of the duplex-CWs (Table 8.6). This may be due to adsorption and precipitation which are more effective in the VF CWs (Zurita and White, 2014). Phosphate removal in CWs during the initial operating period of 1 – 2 years is related to higher availability of phosphorus sorption sites and uptake by plant biomass (Wood et al., 2008). This study showed a < 50% mean phosphate removal efficiency for all the single (VF and HFF) CWs, suggesting saturation of the sorption sites, confirming the PO_4^{3-}-P removal performance of the duplex-CWs treating domestic wastewater (Zapater-Pereyra et al., 2015).

The removal efficiency of the VF CW used in this study is similar to the values reported by Mustapha et al. (2015) for a single VF CW used to treat phoshate in a refinery wastewater with an average influent concentration of 4.0 (± 2.0) mg/L with a removal efficiency of 42%. However, a lower range of 1 – 27% removal efficiency in the HFFs compartment was observed for this study with negative removal at days 7 - 14, 28 and 42 – 49 for the HFF1, while all the measured days in the HFF2 (except for day 42) and day 21 in the HFF3 had a negative removal. This negative removal (higher PO_4^{3-}-P concentration in the effluents) of the HFFs compartment suggests that the HFF wetland acted as a phosphorus source (Wood et al., 2008) and this may reflect possible saturation of the substrate media followed by release from the media (Vohla et al., 2011;Vohla et al., 2017). The HFFs of the duplex-CWs were not planted, thus, no uptake by plants occurred and no additional aeration by plant roots occurred. Thus, P can bind with Fe compounds under the anaerobic conditions due to the reduction of Fe^{3+} to Fe^{2+} and release increased concentrations of phosphates and Fe (Vohla et al., 2007) at the outlet.

P. australis grew very well and rapidly too in all the three VF CWs in the course of this study. This is because PO_4^{3-}-P is readily taken up by plants and immobilized by microbes (Yates, 2008), suggesting that the presence of *P. australis* in the VF CWs could have contributed to the removal of PO_4^{3-}-P in the CWs. This may also have had a positive effect on the phosphate removal performance of the VF CWs by plant uptake. A decrease in the limiting nutrient can inhibit plant growth (Dowty et al. 2001). Thus, the supply of nutrients to the treatment systems in this present study may have aided the *P. australis* growth and survival in the diesel contaminated wastewater. Therefore, the phosphorus removal mechanism by plant uptake may be considered as one of the removal mechanisms. Similarly, adsorption or precipitation by filter media may also have played an important role in phosphorus removal (Ye et al., 2009; Zurita et al., 2014) in this study. Rezaie and Salehzadeh (2014) reported that *P. australis* is able to convey oxygen through its empty space within its tissue from leaves to the stem and the surrounding soils, which is responsible for the decomposition of active aerobic microbial and absorption of pollutants in water systems. Thus, the mechanisms for phosphorus removal in this study may be attributed to plant uptake, adsorption and pecipitation by filter media.

8.4.3. Effect of nutrient addition on duplex-CW performance

This study examined the impact of no nutrients and nutrient supplementation on the performance of duplex-CWs for the treatment of petroleum contaminants. It was observed that in the no nutrient supplement days (the first 20 days), the removal of contaminants in the VF CWs and HFF was high with a higher removal performance in the duplex-CWs (Table 8.4). In contrast, the VF CWs showed a higher performance with no further nutrient removal by the HFF CWs and a corresponding low performance of the duplex-CWs in the nutrient supplemented days (Table 8.5). Previous studies have reported the significance of the addition of nitrogen and phosphorus as well as the supply of oxygen for the biodegradation of petroleum hydrocarbons (Chapelle, 1999; Eke and Scholz, 2008; McIntosh et al., 2017). For instance, Eke and Scholz (2008) reported that the supply of nutrients was more significant for the removal of benzene by VF CW than the role played by *P. australis* or the supply of supplementary oxygen through its rhizomes.

For COD, DRO, benzene, ethylbenzene and o-xylene, 10 mg NH_4^+-N/L and 3 mg PO_4^{3-}-P/L was adequate to produce a high removal performance (>77%), while m/p xylene, toluene and DRO produce a better removal performance (>89%) for higher nutrient (60 mg NH_4^+-N/L and 12 mg PO_4^{3-}-P/L) concentrations. This variability in performance may be due to the duration of the experiment, size of the CW or the type of the hybrid CW (VF CW + HFF). According to Eke and Scholz (2008) excessive nutrient concentrations can inhibit the biodegradation activity with negative effects on biodegradation of hydrocarbons, particularly on the aromatics, suggesting that 60 mg NH_4^+-N/L and 12 mg PO_4^{3-}-P/L may be the reason for the lower contaminants removal by duplex-CW3. Other factors responsible for this variation may also include the type of plant used, the microorganisms present in the CWs, oxygen as well as micronutrients.

Further studies are required to quantify the nutrient and pollutant uptake by plant root biomass to their subsequent translocation to shoot biomass. Further, consideration of the heterogeneity in the study domain under varying (sub)-surface conditions are recommended for future studies to represent more realistic conditions. Also, characterization of the bacterial community and DRO degraders in the duplex-CWs and their role in the remediation process can contribute to further improve the treatment performance of the duplex-CW.

8.4.4 Physical-chemical characteristics of secondary refinery wastewater

The duplex-CWs had mean influent (INF1, INF2 and INF3) pH values that were weakly alkaline and effluents from stage 1 were more acidic than the effluents from stage 2 (Table 8.2). These pH ranges were sufficient for the optimum development of *P. australis* (Calheiros et al., 2007). The pH decrease may be explained by the presence of more H^+ ions, CO_2 production from the decomposing plant litter or other wastewater components trapped in the root zone as well as nitrification of ammonia (Kyambadde et al., 2005). Kyambadde et al. (2005) reported effective nitrification in horizontal surface flow constructed wetland systems between the pH range of 6.5 - 8.6 and temperature range of 5 - 30 ^{0}C.

The influent EC of the duplex-CWs was lower than that of the effluents between days 7 and 28 (Table 8.2) and was not a limiting factor for aquatic plants and microbial growth (Chyan et al., 2013). This was probably due to the addition of ammonium and phosphate to the influent tanks. Also, it may be attributed to evapotranspiration and the effects of evaporation due to the air temperature (> 20 ^{0}C) of the greenhouse (i.e., simulating tropical environment). From day 35 to 56, a lower EC in the effluents was observed. This decrease in the effluent EC after day 28 may be explained by uptake of micro and macroelements by *P. australis* and bacteria as well as adsorption of ions onto plant roots, litter and settleable suspended particles (Kyambadde et al., 2005).

The DO concentrations in the effluent decreased more in duplex-CW 3 (average effluent at HFF = 2.78 mg/L) than in duplex-CW1 and duplex-CW2 (average decrease for each duplex-CW = 3.01 mg/L and 3.19 mg/L, respectively) (Table 8.2). The lower dissolved oxygen measured at stage 2 of the three duplex-CWs is likely attributed to the high oxygen consumption as well as high microbial activity in both stages of this nutrient supplemented duplex-CWs (Pardue et al., 2014; Seeger et al., 2011).

8.5 Conclusion

A series of laboratory and numerical experiments were conducted to investigate the performance of duplex-CWs in treating petroleum contaminants in diesel containing wastewater. The conclusions are:

- The VF CWs component of the duplex-CW had a higher removal efficiency than the HFF CWs, while the duplex-CWs performed better than the VF CW in terms of the removal of the individual petroleum contaminants.

- The duplex-CW showed higher removal efficiencies of the petroleum contaminants in the days with nutrients application than the days without nutrient application.

- This study confirms CWs systems can treat diesel effluents for an integrated secondary treatment, but longer term operation of such units is important to ascertain for the role and benefits of the plants in such systems. The duplex-CW (VF + HFF) systems offer the advantage of obtaining the highest performance levels for the CW technology with a great reduction in size and land requirement.

8.6 Acknowledgement

The authors acknowledge the Government of the Netherlands for their financial assistance via the NUFFIC program (NFP-PhD CF7447/2011). We also thank the IHE laboratory staff for giving the opportunity to conduct this research in their greenhouse and their analytical support

8.7 References

Agarwal, S., Chhibber, V. K., Bhatnagar, A. K., 2013. Review article. Tribological behavior of diesel fuels and the effect of anti-wear additives. Fuel. 106, 21-29. doi:10.1016/j.fuel.2012.10.060

Al-Baldawi, I. A., Abdullah, S. R., Hasan, H. A., Suja, F., Anuar, N., Mushrifah, I., 2014. Optimized conditions for phytoremediation of diesel by *Scirpus grossus* in horizontal subsurface flow constructed wetlands (HSFCWs) using response surface methodology. J Environ Manage. 140, 152 - 159. doi:10.1016/j.jenvman.2014.03.007

Al-Baldawi, I. A., Abdullah, S. R., Suja, F., Anuar, N., Mushrifah, I., 2013. Comparative performance of free surface and sub-surface flow systems in the phytoremediation of hydrocarbons using *Scirpus grossus*. J. Environ. Manage. 130, 324-330. doi:10.1016/j.jenvman.2013.09.010

APHA, 2002. Standard Methods for the Examination of Water and Wastewater, twentieth ed. Baltimore, Maryland, USA. American Public Health Association.

Aslam, M. M., Malik, M., Baig, M., Qazi, I., Iqbal, J., 2007. Treatment performances of compost-based and gravel-based vertical flow wetlands operated identically for refinery wastewater treatment in Pakistan. Ecol. Eng. 30, 34–42. doi:10.1016/j.ecoleng.2007.01.002

Ávila, C., Matamoros, V., Reyes-Contreras, C., Piña, B., Casado, M., Mita, L., Rivetti, C., Barata, C., García, J., Bayona, J. M., 2014. Attenuation of emerging organic contaminants in a hybrid constructed wetland system under different hydraulic loading rates and their associated toxicological effects in wastewater. Sci. Total Environ. 470–471, 1272–1280. doi:10.1016/j.scitotenv.2013.10.065

Balachandran, C., Duraipandiyan, V., Balakrishna, K., Ignacimuthu, S., 2012. Petroleum and polycyclic aromatic hydrocarbons (PAHs) degradation and naphthalene metabolism in *Streptomyces sp.* (ERI-CPDA-1) isolated from oil contaminated soil. Bioresour. Technol. 112, 83-90. doi:10.1016/j.biortech.2012.02.059

Ballesteros Jr., F., Vuong, T. H., Secondes, M. F., Tuan, P. D., 2016. Removal efficiencies of constructed wetland and efficacy of plant on treating benzene. Sustainable Environ. Res. 26, 93-96. doi:10.1016/j.serj.2015.10.002

Basu, S., K., Y. B., Mathur, S., 2015. Enhanced bioremediation of BTEX contaminated groundwater in pilot-scale wetlands. Environ. Sci. Pollut. Res. 22, 20041–20049. doi:10.1007/s11356-015-5240-x.

Burland, S. M., Edwards, E. A., 1999. Anaerobic benzene biodegradation linked to nitrate reduction. Appl. Environ. Microbiol. 65, 529-533.

Calheiros, C., Rangel, A., Castro, P., 2007. Constructed wetland systems vegetated with different plants applied to the treatment of tannery wastewater. Water Res. 41, 1790-1798. doi:10.1016/j.watres.2007.01.012

Chapelle, F., 1999. Bioremediation of petroleum hydrocarbon-contaminated ground water: The perspectives of history and hydrology. Ground Water. 37, 122-132. doi:10.1111/j.1745-6584.1999.tb00965.x

Chavan, A., Mukherji, S., 2008. Treatment of hydrocarbon-rich wastewater using oil degrading bacteria and phototrophic microorganisms in rotating biological contactor: Effect of N:P ratio. J. Hazard Mater. 154, 63–72. doi:10.1016/j.jhazmat.2007.09.106

Chen, Z., Kuschk, P., Reiche, N., Borsdorf, H., Kästner, M., Köser, H., 2012. Comparative evaluation of pilot scale horizontal subsurface-flow constructed wetlands and plant root mats for treating groundwater contaminated with benzene and MTBE. J. Hazard Mater. 209-210. doi:10.1016/j.jhazmat.2012.01.067

Chyan, J.-M., Senoro, D.-B., Lin, C.-J., Chen, P.-J., Chen, M.-L., 2013. A novel biofilm carrier for pollutant removal in a constructed wetland based on waste rubber tire chips. Int. Biodeterior. Biodegrad. 85, 638-645. https://doi.org/10.1016/j.ibiod.2013.04.010

Dittmer, U., Meyer, D., Langergraber, G., 2005. Simulation of a subsurface vertical flow constructed wetland for CSO treatment. Water Sci. Technol. 51, 225-232.

Dowty, R. A., Shaffer, G. P., Hester, M. W., Childers, G. W., Campo, F. M., Greene, M. C., 2001. Phytoremediation of small-scale oil spills in fresh marsh environments: a mesocosm simulation. Marine Environ. Res. 52, 195-211. doi:10.1016/S0141-1136(00)00268-3

Drizo, A., Frost, C. A., Grace, J., Smith, K. A., 2000. Phosphate and ammonium distribution in a pilot-scale constructed wetland with horizontal subsurface flow using shale as a substrate. Water Res. 34, 2483-2490.

Eckert, P., Appelo, C. A., 2002. Hydrogeochemical modeling of enhanced benzene, toluene, ethylbenzene, xylene (BTEX) remediation with nitrate. Water Resourc. Res. 38, 5-11. doi:10.1029/2001WR000692

Eke, P. E., Scholz, M., 2008. Benzene removal with vertical - flow constructed treatment wetlands. J. Chem. Technol. Biotechnol. 83, 55-63. doi:10.1002/jctb.1778

Farhadian, M., Duchez, D., Vachelard, C., Larroche, C., 2008. Monoaromatics removal from polluted water through bioreactors - A review. Water Res. 42, 1325-1341. doi:10.1016/j.watres.2007.10.021

Gupta, P. K., Yadav, B. K., 2017. Bioremediation of Non-aqueous Phase Liquids (NAPLS) Polluted Soil and Water Resources. In: Environmental Pollutants and their Bioremediation Approaches. Florida, USA. CRC Press, Taylor and Francis Group.

Gupta, P. K., Ranjan, S., Yadav.B.K., 2013. BTEX biodegradation in soil-water system having different substrate concentrations. Int. J. Eng. Res. Technol. (IJERT), 2, 1765-1772.

Henrichs, M., Langergraber, G., Uhl, M., 2007. Modelling of organic matter degradation in constructed wetlands for treatment of combined sewer overflow. Sci. Total Environ. 380, 196-209. doi:10.1016/j.scitotenv.2006.11.044

Herrera Melián, J., Rodríguez, A. M., Arãna, J., Díaz, O. G., Henríquez, J. G., 2010. Hybrid constructed wetlands for wastewater treatment and reuse in the Canary Islands. Ecol. Eng. 36, 891–899. doi:10.1016/j.ecoleng.2010.03.009

Hijosa-Valsero, M., Matamoros, V., Pedescoll, A., Martín-Villacorta, J., Bécares, E., Garcıá, J., Bayona, J. M., 2011. Evaluation of primary treatment and loading regimes in the removal of pharmaceuticals and personal care products from urban wastewaters by subsurface-flow constructed wetlands. Int. J. Environ. Anal. Chem. 91, 632-653. doi:10.1080/03067319.2010.526208

Kyambadde, J., Kansiime, F., Dalhammar, G., 2005. Nitrogen and phosphorus removal in the substrate-free pilot constructed wetlands with horizontal surface flow in Uganda. Water Air Soil Pollut. 165, 37-59.

Langergraber, G., Šimůnek, J., 2005. Modeling variably saturated water flow and multicomponent reactive transport in constructed wetlands. Vadoze Zone. 4, 924-938.

Langergraber, G., Šimůnek, J., 2006. The Multi-Component Reactive Transport Module CW2D for Constructed Wetlands for the HYDRUS Software Package. Dep. of Environ. Sciences, California.

Liang, F., Lu, M., Keener, T. C., Liu, Z., Khang, S.-J., 2005. The organic composition of diesel particulate matter, diesel fuel and engine oil of a non-road diesel generator. J. Environ. Monit. 7, 983 – 988. doi:10.1039/b504728e

Małoszewski, P., Wachniew, P., Czupryński, P., 2006. Study of hydraulic parameters in heterogeneous gravel beds: Constructed wetland in Nowa Słupia (Poland). J. Hydrol. 331, 630-642. doi:10.1016/j.jhydrol.2006.06.014

Masi, F., Martinuzzi, N., 2007. Constructed wetlands for the mediterranean countries: hybrid systems for water reuse and sustainable sanitation. Desalination. 215, 44-55. doi:10.1016/j.desal.2006.11.014

Mathur, S., Yadav, B. K., 2009. Phytoextraction modeling of heavy metal (lead) contaminated site using maize (*Zea mays*). Practice Periodical of Hazardous Toxic and Radioactive Waste Management. 13, 229-238.

Mayo, A., Bigambo, T., 2005. Nitrogen transformation in horizontal subsurface flow constructed wetlands: I. Model development. Phys. Chem. Earth Parts A/B/C. 30, 658-667. doi:10.1016/j.pce.2005.08.005

McIntosh, P., Schulthess, C. P., Kuzovkina, Y. A., Guillard, K., 2017. Bioremediation and phytoremediation of total petroleum hydrocarbons (TPH) under various conditions. Int. J. Phytorem. 19, 755-764. http://dx.doi.org/10.1080/15226514.2017.1284753

Meyer, D., Chazarenc, F., Claveau-Mallet, D., Dittmer, U., Forquet, N., Molle, P., Morvannou, A., Pálfy, T., Petitjean, A., Rizzo, A., Samsó Campà, R., Scholz, M., Soric, A., Langergraber, G., 2015. Modelling constructed wetlands: scopes and aims - a comparative review. Ecol. Eng. 80, 205-213.

Mustapha, H. I., Rousseau, D., van Bruggen, J., Lens, P., 2011. Treatment performance of horizontal subsurface flow constructed wetlands treating inorganic pollutants in simulated refinery effluent. In:2nd Biennial Engineering Conference. School of Engineering and Engineering Technology, Federal University of Technology, Minna.

Mustapha, H. I., van Bruggen, J. J., Lens, P. N., 2015. Vertical subsurface flow constructed wetlands for polishing secondary Kaduna refinery wastewater in Nigeria. Ecol. Eng. 84, 588-595. doi:10.1016/j.ecoleng.2015.09.060

Pardue, M. J., Castle, J. W., Rodgers Jr., J. H., Huddleston III, G. M., 2014. Treatment of oil and grease in produced water by a pilot-scale constructed wetland system using biogeochemical processes. Chemosphere. 103, 67–73. doi:10.1016/j.chemosphere.2013.11.027

Pastor, R., Benqlilou, C., Paz, D., Cardenas, G., Espuna, A., Puigjaner, L., 2003. Design optimisation of constructed wetlands for wastewater treatment. Resour. Conserv. Recycling. 37, 193-204. doi:10.1016/S0921-3449(02)00099-X

Pawlak, Z., Rauckyte, T., Oloyede, A., 2008. Oil, grease and used petroleum oil management and environmental economic issues. J. Achievements Mater. Manufact. Eng. 26, 11-17.

Rezaie, H., Salehzadeh, M., 2014. Performance removal nitrate and phosphate from treated municipal wastewater using *Phragmites australis* and *Typha latifolia* aquatic plants. J. Civil Eng. Urbanism. 4, 315-321.

Ribeiro, H., Almeida, C. M., Mucha, A. P., Teixeira, C., Bordalo, A. A., 2013. Influence of natural rhizosediments characteristics on hydrocarbons degradation potential of microorganisms associated to *Juncus maritimus* roots. Int. Biodeterioration Biodegrad. 84. doi:0.1016/j.ibiod.2012.05.039

Ribeiro, H., Mucha, A. P., Almeida, M. R., Bordalo, A. A., 2013. Bacterial community response to petroleum contamination and nutrient addition in sediments from a temperate salt marsh. Sci. Total Environ. 458-490, 568-576. doi:10.1016/j.scitotenv.2013.04.015

Rizzo, A., Langergraber, G., 2016. Novel insights on the response of horizontal flow constructed wetlands to sudden changes of influent organic load: A modeling study. Ecol. Eng. 93, 242-249.

Saeed, T., Guangzhi, S., 2012. A review on nitrogen and organics removal mechanisms in subsurface flow constructeds: Dependency on environmental parameters, operating conditions and supporting media. J. Environ. Manage. 112, 429-448. doi:10.1016/j.jenvman.2012.08.011

Saien, J., Shahrezaei, F., 2012. Organic pollutants removal from petroleum refinery wastewater with nanotitania photocatalyst and UV light emission. Int. J. Photoenergy. 1-5. doi:10.1155/2012/703074

Seeger, E. M., Kuschk, P., Fazekas, H., Grathwohl, P., Kaestner, M., 2011. Bioremediation of benzene-, MTBE- and ammonia-contaminated groundwater with pilot-scale constructed wetlands. Environ. Pollut. 199, 3769-3776. doi:10.1016/j.envpol.2011.07.019

Semrany, S., Favier, L., Djelal, H., Tahac, S., Amrane, A., 2012. Review. Bioaugmentation: possible solution in the treatment of bio-refractory organic compounds (Bio-ROCs). Biochem. Eng. J. 69, 75-86. doi:10.1016/j.bej.2012.08.017

Šimůnek, J., van Genuchten, M., Šejna, M., 2006. The HYDRUS software package for simulating the two- and three-dimensional movement of water, heat, and multiple solutes in variably-saturated media. Version 1.0. PC-Progress, 1.0, 1-241. Prague, Prague, Czec republic.

Tobiszewski, M., Tsakovski, S., Simeonov, V., Namies'nik, J., 2012. Chlorinated solvents in a petrochemical wastewater treatment plant: An assessment of their removal using self-organising maps. Chemosphere. 87, 962-968. doi:10.1016/j.chemosphere.2012.01.057

Tomenko, V., Ahmed, S., Popov, V., 2007. Modelling constructed wetland treatment system performance. Ecol. Modell. 205, 355-364. doi:10.1016/j.ecolmodel.2007.02.030

Toscano, A., Langergraber, G., Consoli, S., Cirelli, G., 2009. Modelling of pollutant removal in a pilot-scale two-stage subsurface flow constructed wetlands. Ecol. Eng. 35, 281-289.

van Afferden, M., Rahman, K. Z., Mosig, P., De Biase, C., Thullner, M., Oswald, S. E., Müller, R. A., 2011. Remediation of groundwater contaminated with MTBE and benzene: The potential of vertical-flow soil filter systems. Water Res. 45, 3053-3074. doi:10.1016/j.watres.2011.07.010

van Genuchten, M. T., 1980. A closed-form equation for predicting the hydraulic conductivity of unsaturated soils. Soil Sci. Soc. Am. J. 44, 892–898.

Vohla, C., Alas, R., Nurk, K., Baatz, S., Mander, U., 2007. Dynamics of phosphorus, nitrogen and carbon removal in a horizontal subsurface flow constructed wetland. Sci. Total Environ. 380, 66-74. doi:10.1016/j.scitotenv.2006.09.012

Vohla, C., Koiv, M., Bavor, J., Chazarenc, F., Mander, U., 2011. Filter materials for phosphorus removal from wastewater in treatment wetlands - A review. Ecol. Eng. 37, 70 - 89. doi:10.1016/j.ecoleng.2009.08.003

Vymazal, J., 2005. Horizontal sub-surface flow and hybrid constructed wetlands systems for wastewater treatment. Ecol. Eng. 25, 478-490. doi:10.1016/j.ecoleng.2005.07.010

Wake, H., 2005. Oil refineries: a review of their ecological impacts on the aquatic environment. Estuarine Coastal Shelf Sci. 62, 131-140. doi:10.1016/j.ecss.2004.08.013

Wallace, S., Schmidt, M., Larson, E., 2011. Long term hydrocarbon removal using treatment wetlands. SPE Annual Technical Conference and Exhibition. pp. 1-10. Denver, Colorado. Society of Petroleum Engineers (SPE) International.

Wood, J. D., Gordon, R., Madani, A., Stratton, G., 2008. A long term assessment of phosphorus treatment by a constructed wetland receiving dairy wastewater. Wetlands. 28, 715-723.

Wu, S., Wallace, S., Brix, H., Kuschk, P., Kirui, W. K., Masi, F. D., 2015. Treatment of industrial effluents in constructed wetlands: Challenges, operational strategies and overall performance. Environ. Pollut. 201, 107-120. doi:10.1016/j.envpol.2015.03.006

Wynn, M., Liehr, S., 2001. Development of a constructed subsurface flow wetland simulation model. Ecol. Eng. 16, 519-536. doi:10.1016/S0925-8574(00)00115-4

Yadav, B. K., Hassanizadeh, S. M., 2011. An overview of biodegradation of LNAPLs in coastal (semi)-arid environment. Water Air Soil Pollut. 220, 225-239. doi: 10.1007/s11270-011-0749-1

Yadav, B. K., Ansari, F. A., Basu, S., Mathur, A., 2014. Remediation of LNAPL contaminated groundwater using plant-assisted biostimulation and bioaugmentation methods. Water Air Soil Pollut. 225, 1793-. doi:10.1007 /s11270-013-1793-9

Yadav, B. K., Mathur, S., Siebel, M. A., 2009. Soil moisture flow modeling with water uptake by plants (wheat) under varying soil and moisture conditions. J. Irrigation Drainage Eng. 135, 375 - 381.

Yang, L., Hu, C, 2005. Treatments of oil-refinery and steel-mill wastewaters by mesocosm constructed wetland systems. Water Sci. Technol. 51, 157-164.

Yates, C., 2008. Comparison of Two Constructed Wetland Substrates for Reducing Phosphorus and Nitrogen Pollution in Agricultural Runoff. Department of Bioresource Engineering, Faculty of Agricultural and Environmental Sciences. McGill University, Montreal

Ye, F., Li, Y., 2009. Enhancement of nitrogen removal in towery hybrid constructed wetland to treat domestic wastewater for small rural communities. J. Ecol Eng, 35, 1043-1050. doi:doi:10.1016/j.ecoleng.2009.03.009

Zapater-Pereyra, M., Ilyas, H., S, L., van Bruggen, J., Lens, P., 2015. Evaluation of the performance and the space requirement by three different hybrid constructed wetlands in a stack arrangement. Ecol. Eng. 82, 290-300.

Zurita, F., White, J. R., 2014. Comparative study of three two-stage hybrid ecological wastewater treatment systems for producing high nutrient, reclaimed water for irrigation reuse in developing countries. Water. 6, 213-228. doi:10.3390/w6020213

Chapter 9. General discussion and outlook

9.1. Introduction

In temperate and developed countries, research and applications of constructed wetlands (CWs) have been implemented extensively for over the last three decades (Vymazal, 2011). This is not the case in (sub)tropical and developing countries (Kivaisi, 2001). Information on design, operation and maintenance of CWs treating petroleum refining wastewater under tropical climatic conditions is scarce. This PhD study stands as the first documented study in Nigeria for this type of wastewater under tropical climatic conditions.

The focus of this PhD thesis was on the use of subsurface flow (SSF) and hybrid CWs for polishing secondary treated petroleum refinery wastewater. To achieve this aim, five sets of experiments were conducted as illustrated in Fig. 9.1 based on the specific objectives (Chapter 1; Table 9.1). The first part of this research was a preliminary study to investigate the potentials of treating refinery wastewater in aerated and non-aerated horizontal subsurface flow constructed wetlands planted with *Phragmites australis* as well as to examine the effectiveness of increasing aeration on the removal of contaminants (Mustapha et al., 2011). The results of the analysis revealed a higher removal efficiency in the aerated horizontal subsurface flow constructed wetlands than in the non-aerated horizontal subsurface flow constructed wetlands (Mustapha et al., 2011) for all the monitored parameters (chemical oxygen demand (COD), biological oxygen demand (BOD), total nitrogen, cadmium (Cd), chromium (Cr), lead (Pb) and zinc (Zn)).

The other phases of the research (Fig. 9.1) include the characterization of secondary treated refinery wastewater (Chapter 3), the design, construction and operation of hybrid (vertical and horizontal) subsurface flow constructed wetlands for the polishing of secondary treated refinery wastewater (Chapters 4 and 5). The research goal for Chapter 5 is the optimization of vertical subsurface flow (VSSF) CWs for secondary treated refinery wastewater and Chapter 6 investigated the fate of heavy metals in these CWs. Based on the findings of Chapter 4, 5 and 6, the research goal for Chapter 7 was the choice of *T. latifolia* plant species and vertical subsurface flow constructed wetlands for the treatment of benzene, toluene, ethylbenzene and xylene (BTEX), phenol and oil and grease (O&G) in the secondary treated refinery effluent. Finally, the performance of duplex constructed wetlands (duplex-CWs) was assessed for the treatment of diesel range organics in simulated refinery effluent (Chapter 8).

The outcome of these studies is discussed in this chapter. In addition, the types of macrophytes used, the implications for the application of CWs for the Kaduna refining and petrochemical company (KRPC) (Kaduna, Nigeria), scaling up of the experimental design to cater for the effluent discharged by KRPC (Table 9.2) and recommendations for future research are discussed.

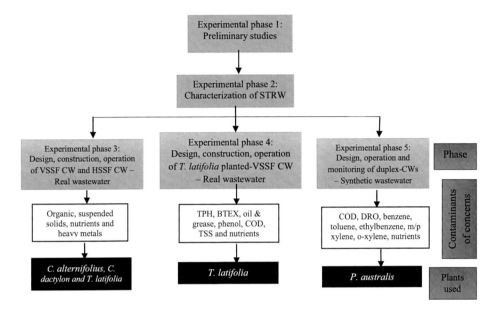

Fig. 9.1. Schematic diagram showing the experimental phases, contaminants of concerns and the types of macrophytes used in the polishing of secondary treated petroleum refinery wastewater. Note: STRW- secondary treated petroleum refinery wastewater; TPH – total petroleum hydrocarbon, BTEX - benzene, toluene, ethylbenzene, xylene, COD – chemical oxygen demand, TSS – total suspended solids, DRO – diesel range organics.

9.2. Petroleum refinery wastewater characterization

The processes of refining crude oil into various products require large volumes of water. The KRPC Nigeria discharges approximately 100,000 m³ of secondary treated refinery wastewater per day during the dry season, with an additional 14,000 m³/day from rainfall during the wet season. Information about the generated wastewater is presented in Table 9.2. The secondary treated refinery wastewater from KRPC was characterized during a period of sixteen months to have baseline information on the quality of the wastewater discharged by the refinery. The characterized secondary treated refinery wastewater was laden with organic and inorganic contaminants with a high variation in composition (Chapter 3). This is due to the differences in the processes of refining crude oil (Wake, 2005).

This study is based on selected contaminants of concern: organic matter, suspended solids, heavy metals, nitrogen, phosphorus, phenol and O&G. The composition of secondary treated refinery wastewater ranged from 6.5 – 8.5 (pH), 22 – 31 ^{0}C (temperature), 1113 – 1957 µS/cm (electrical conductivity), 101 - 283 mg/L BOD, 135 – 521 mg/L COD, 0.00 - 1.16 µg/L phenol, 0.0 – 14.2 mg/L O&G, 0.30 – 12.5 mg/L ammonium-N, 1.32 – 15.64 mg/L phosphate-P, 0.0 – 0.03 mg/L Cd, 0.0 – 3.4 mg/L Cr, 0.0 – 0.05 mg/L Cu, 0.03 – 0.80 mg/L Zn, 0.49 – 16.9 mg/L Fe and 0.0 – 0.06 mg/L Pb. The secondary treated refinery wastewater was compared with national (Federal Environmental Protection Agency (FEPA), Nigeria) and international (WHO and EU) standards to confirm their compliance. The bases for the sixteen months analysis was to enable the establishment of a trend for the variability in the concentrations of the monitored parameters. Thus, the monitored parameters were subjected to rank-based non-parametric Mann-Kendall tests (a spreadsheet Makesen 1.0) to establish a statistical significant trend in the varied concentrations. The results showed that only 6 of the 23 parameters depict significant trends. The six parameters are dissolved oxygen (DO), total dissolved solids (TDS), total solids (TS), COD, Cr and Fe. The trends for these six parameters were negatively significant, implying reductions of the concentration of the parameters during the period under consideration. In addition, the Cd and Pb concentration showed no statistical significant trend.

Despite the wastewater treatment processes (preliminary, primary and secondary) carried out by KRPC (Nigeria) before discharge into the environment, it is not compliant with FEPA, EU and WHO guideline values for wastewater discharge limits. The secondary treated refinery wastewater did not meet the FEPA (Nigeria) requirement for electrical conductivity, BOD, COD, Cr, Cd, phenol as well as O&G in wastewater to be discharged into the environment. Thus, further treatment by low-cost technology is required to clean the wastewater so that it can be safely disposed into the environment.

Constructed wetlands have been effectively used to treat petroleum contaminated wastewater in the developed world both at large (Wallace et al., 2011; Davis et al., 2009; Simi, 2000) and small (Horner et al., 2012; Ji et al., 2007) scales to safe limits. Regrettably, the technology has been largely ignored in developing countries where effective, low cost wastewater treatment strategies are critically needed (Dipu and Thanga, 2009; Kivaisi, 2001). Thus, installation of CWs at the effluent discharge point of the refinery for polishing would prevent the discharge of contaminants that exceeds the water quality standards on a cumulative basis, thus improving the quality of the wastewater discharged into the Romi River.

Table 9.1: Summary of the thesis chapters, experiments, experimental site locations, important results and conclusions

Experiment (chapter)	Description	Aim/objective	Experimental design	Most important results	Main conclusion	Location of experimental site	Publication
Phase 1 (PhD proposal)	Preliminary studies	To examine the effects of aerated and non-aerated horizontal subsurface flow (HSSF) CW on reduction of constituents of concern	Duration: 122 days; Wastewater: simulated refinery effluent; CW type: HSSF; HRT: 2 d; HLR: 11 L/d; Flow mode: continuous; Plant: *Phragmites australis*	There were no large variations in the removal efficiency between the aerated and the non-aerated treatments systems. The aerated treatment had a higher mean removal percentage (51.3 ± 31.2 - 94.2 ± 7.7 %) than the non-aerated wetland (49.7 ± 29.7 - 93.8 ± 7.5 %) for COD, BOD, TN, Cd, Cr, Pb and Zn.	The HSSF CWs was capable of removing organic and inorganic pollutants from the simulated refinery effluent.	Laboratory, UNESCO-IHE, Delft, the Netherlands	Mustapha et al., 2011. *In:* Book of Proceedings
Phase 2 (Chapter 3)	Physiochemical analysis	To have a baseline (background) information on the quality of wastewater discharged by KRPC into the environment	Duration: 487 days; Wastewater: real Analysis: rank-based non-parametric Mann–Kendall (M-K) test	TSS (86.1 ± 99.7 mg/L), BOD (106 ± 58.9 mg/L), COD (232 ± 121.2 mg/L), turbidity (56.8 ± 59.2 NTU), ammonium (3.6 ± 4.4 mg/L) and phosphate (2.3 ± 1.5 mg/L) were above the effluent permissible	Background information on the nature of wastewater was generated; 6 out of the 23 parameters monitored showed significant trends; Further treatment is needed before	KRPC field site, Kaduna, Nigeria	Mustapha et al., 2015. Journal of Ecological Engineering

Phase (Chapter)	Focus	Aim	Methods	Results		Location	Reference
Phase 3 (Chapter 4)	Design, experimental setup, plant establishment, operation and monitoring of SSF CWs	To assess the potentials of subsurface flow constructed wetlands for treatment of secondary refinery effluent in Nigeria	Duration: 308 days; Wastewater: real CW type: VSSF; HRT: 48 h; HLR: 4.65 L/m²h; Flow mode: continuous; Plant: *Cyperus alternifolius* and *Cynodon dactylon* (L.) *Pers.*	Turbidity: 5.4 (± 5.2) and 5.9 (± 7.3) NTU; BOD: 31.0 (± 32.9) and 28.7 (± 31.4); COD: 57.3 (± 71.6) mg/L and 78.8 (± 57.7) mg/L; ammonium (0.6 (± 0.7) and 0.9 (± 1.3) mg/L and phosphate (2.3 (± 1.5) and 2.4 (± 1.7) mg/L), respectively for *C. alternifolius* and *C. dactylon*	discharge limits (WHO and FEPA). the effluent can be discharged into the environment. Effluent concentrations were reduced in the *C. alternifolius* and *C. dactylon* planted wetland VSF CWs. *C. alternifolius* and *C. dactylon* planted VSF CWs were not significantly different from each other.	KRPC field site, Kaduna, Nigeria	Mustapha et al., 2015. Journal of Ecological Engineering
Phase 3 (Chapter 5)	Hybrid constructed wetlands	To evaluate the removal efficiency of *Typha latifolia* planted VSSF CWs and coupling HSSF CWs to the VSSF CW systems	Duration: 308 days; Wastewater: real CW type: VSSF + HSSF; VSSF CW - HRT: 48 h; HLR: 4.65 L/m²h; HSSF CW - HRT: 148 h; HLR: 1.08 L/m²h Plant: *T.*	The VSSF CWs had a removal efficiency of 76% for BOD, 73% for COD, 70% for ammonium-N, 68% for nitrate-N, 49% for phosphate, 68% for TDS and 89% for turbidity; The hybrid CW showed an improved effluent quality with removal efficiencies of,	In comparison to *C. alternifolius* and *C. dactylon* planted VSSF CWs, *T. latifolia* planted VSSF CW achieved better removal efficiencies for all parameters measured; The *T. latifolia* planted hybrid CW had final effluent concentrations	KRPC field site, Kaduna, Nigeria	Mustapha et al., 2018. Journal of Water Air & Soil Pollution

				latifolia	respectively, 94% for BOD, 88% for COD, 84% for ammonium-N, 89% for nitrate-N, 78% for phosphate-P, 85% for TSS and 97% for turbidity.	that conform very well to the discharge limits of WHO and FEPA (Nigeria).	
Phase 3 (Chapter 6)	Heavy metals	To examine the performance of pilot scale VSF-CW in removing heavy metals from secondary treated refinery wastewater under tropical conditions	Duration: 308 days; Wastewater: real CW type: VSSF; HRT: 48 h; HLR: 4.65 L/m²h; Flow mode: continuous; Plant: *Typha latifolia, Cyperus alternifolius* and *Cynodon dactylon*	The *T. latifolia* planted VSF-CW had the best heavy metal removal performance, followed by the *C. alternifolius* planted VSF CW and then the *C. dactylon* planted VSF-CW; Cu, Cr, Zn, Pb, Cd and Fe were accumulated in the plants at all the three VSF-CWs; the accumulation of the heavy metals in the plants accounted for only a rather small fraction (0.09 - 16 %) of the overall heavy metal removal by the wetlands.	*T. latifolia, C. alternifolius* and *C. dactylon* were able to efficiently remove Cd, Cr, Cu, Zn, Fe and Pb from the secondary treated refinery wastewater; The three plants were able to tolerate Cd, Cr, Cu, Zn, Fe and Pb higher than their threshold limits without showing any sign of toxicity. The plants did not accumulate sufficient quantities of heavy metals to qualify them as hyper accumulators.	KRPC field site, Kaduna, Nigeria	Mustapha et al., 2018. International Journal of Phytoremediation

Phase	Topic	Objective	Details	Results	Location	Reference
Phase 4 (Chapter 7)	Organic pollutants	To examine the potentials of the VSF CWs to remove TPH, phenol and oil and grease from secondary refinery wastewater	Duration: 181 days; Wastewater: real CW type: VSSF; HLR: 3.2 L/m²h; Flow mode: continuous; Plant: *Typha latifolia* Substrate media spiked with cow dung	The *T. latifolia* planted VSF CWs removed 70 % TPH, 99 % phenol, 75 % O&G, 70 % COD and 69 % TSS	The VSF CWs showed effective removal of TPH, phenol, O&G, TN and TP; *T. latifolia* was able to accumulate these contaminants into its plant tissues and showed higher translocation ability for TPH, phenol, O&G and TN with the exception of TP that was mainly retained in *T. latifolia* roots.	KRPC field site, Kaduna, Nigeria	Submitted to a Journal for possible evaluation.
Phase 5 (Chapter 8)	Duplex-CW	To evaluate the potential of the hybrid CW system to treat influent wastewater containing diesel range organic compounds under controlled conditions in a greenhouse using	Duration: 117 days; Wastewater: simulated CW type: VSSF + HFF; VSSF CW - HRT: 4.8 d; HLR: 4.65 L/m²h; HSSF CW - HRT: 2.1 d; HLR: 0.066 m³/m²h; Flow mode: continuous;	The VSSF CWs had a removal efficiency of 74 %; 64% and 69 % for COD; 86 %, 90 % and 93 % for benzene, 97 %, 65 % and 54 % for ammonium-N, 40 %, 42 % and 50 % for phosphate, respectively for VFCW 1, VFCW 2 and VFCW 3; The hybrid CW	The three duplex-CWs had higher removal efficiencies of DRO, benzene, toluene, ethylbenzene, m/p-xylene and o-xylene in the period of nutrients addition (days 21 to 56) compared with the period without nutrients	Greenhouse, UNESCO-IHE, Delft, the Netherlands	Mustapha et al., 2018. Journal of Chemosphere

| laboratory and numerical experiments | Plant: *P. australis* | showed an improved effluent quality with removal efficiencies of, 100 %, 100 % and 97 % DRO; 100 %, 98 % and 89 % ethylbenzene; for duplex-CW 1, 2 and 3 with nutrient supplement. | (first 20 days); Comparing the performance of the single CWs to the hybrid system, the VF CWs showed higher removal efficiency for COD and benzene, whereas, the hybrid system showed a higher removal efficiency for toluene, ethylbenzene, m/p xylene and o-xylene. |

Table 9.2. Wastewater generated by Kaduna refinery and petrochemical company

S/N	Wastewater	Flow rate (m³/h)	Source
1.	Process oily wastewater:	Max. 88	On-site
	Sour water stripping/desalter	55	
	Fluid catalytic cracking sour water stripping	15.8	
	Lube plant	11.0	
	Spent caustic treater	2.0 (batch)	
	Boiler blowdown/laboratory	4.2	
2.	Sanitary waste	15	Lavatory
3.	Oily water	196	Pump station
	Pump cooling water (oily rain water)	3000	On-site tank farm
4.	Cooling water blowdown	50	Cooling tower
	Demineralizer neutralized waste		demineralizer
5.	Rainwater	14,000	Whole refinery
6.	Clean water	50	
7.	Raw water treatment clarifier	20	72so 1AB – clarifier

*Source: Kaduna refinery and petrochemical company.

9.3. Wastewater treatment by constructed wetlands

CWs have a low energy requirement, are environmentally friendly and are natural treatment systems. Other advantages of CWs are the less intrusive approach, low maintenance and solar-driven biological processes (Chapter 2). CW systems protect resources for future generations by preserving the ecosystem as well as protecting biodiversity (Schröder et al., 2007). They are also economically viable (Shpiner et al., 2009), especially for developing countries with limited water resources and means (Mustapha et al., 2015). Therefore, they can be used as alternative to conventional wastewater treatment or as tertiary treatment for improved final wastewater quality.

Surface flow (SF), vertical subsurface flow (VSSF) (Chapter 4), horizontal subsurface flow (HSSF) (Chapter 5) and hybrid CWs (Chapter 8) have been successfully used for the treatment of petroleum refinery wastewater. This chapter gives the main findings on the performance of the pilot scale VSSF, HSSF and hybrid (VSSF + HSSF CWs and duplex

CWs), the implications of discharging secondary treated wastewater into the Romi River and the design criteria for a full-scale facility placed at the effluent discharge point. Thus, the outlook on the opportunities for the use of and research into CWs for refinery wastewater treatment is presented.

9.3.1. Vertical subsurface flow constructed wetlands (VSSF CWs)

The potential of CWs for the removal of BOD_5, COD, suspended solids and nutrients from domestic wastewater is well documented (Yadav et al., 2012). However, there are only few studies on petroleum-contaminated wastewater. Thus, VSSF CWs planted with *C. alternifolius* and *C. dactylon* were used to investigate the removal of BOD_5, COD, TSS, TDS, TS, turbidity, ammonium-N and phosphate-P from secondary treated refinery wastewater (Chapter 4). The VSSF CW showed a removal efficiency within the range of 43 – 85% by the planted *C. alternifolius* and a range of 42 – 82% by *C. dactylon*. Both planted VSSF CWs were not significantly different from each other. However, both planted VSSF CWs exceeded the FEPA and WHO guideline values for BOD_5, COD, and ammonium-N (Chapter 4). This may indicate that the oxygen released from the roots was less than the amount required for the aerobic decomposition of BOD_5 and COD as well as the nitrification of NH_4^+-N in the *C. alternifolius* and *C. dactylon* planted VSSF CWs (Vymazal, 2011; Abdulhakeem et al., 2016).

For an improved performance of the treatment system, Abdulhakeem et al. (2016) suggested artificially aerating the treatment systems and increasing their hydraulic retention time. Also, appropriate choice of macrophytes can influence treatment processes (Calheiros et al., 2007; Liu et al., 2007; Madera-Parra et al., 2015). Accordingly, *T. latifolia*-planted VSSF CWs were assessed for their performance for the removal of these contaminants (Chapter 5). In comparison to *C. alternifolius* and *C. dactylon* planted VSSF CWs, the *T. latifolia*-planted VSSF CW achieved a better removal efficiency for all the parameters measured and the polished wastewater met the discharge limits (Chapter 5). Thus, the extensive root system of *T. latifolia* (Fig. 9.2) might have enhanced the VSSF CWs removal efficiency by providing larger substrate for bacterial attachment and oxygenation (Vymazal, 2011). *T. latifolia* may have been more tolerant of the pollutant loadings as well as have a higher growth rate compared to the other plant species investigated (Chapters 5, 6 and 7).

CWs have demonstrated the effectiveness of heavy metal removal from petroleum refinery wastewater (Aslam et al., 2010; Mustapha et al., 2011; Horner et al., 2012; Chapter 6). *T.*

latifolia, *C. alternifolius* and *C. dactylon* planted VSSF CWs were used to remove Cu, Cr, Zn, Pb, Cd and Fe from secondary treated refinery wastewater under tropical conditions (Chapter 6; Mustapha et al., 2018). *T. latifolia*-planted VSSF CW had a range of removal efficiencies from 74 to 95 %, *C. alternifolius*-planted VSSF CW had a removal efficiency from 66 to 90 % and *C. dactylon*-planted VSSF CW had a removal efficiency from 61 to 90%. The results showed variability from plant species to plant species, suggesting that the different planted systems have different capabilities for metal removal (Chapter 6; Mustapha et al., 2018). Cu, Cr, Zn, Pb, Cd and Fe were taken up into the root, leaf and stem parts of the plants, with the roots being the most significant (Chapter 6). The three plant species showed high metal tolerance by accumulating metals higher than their phytotoxic limits (Chapter 6; Nouri et al., 2009). However, none of the plant species was able to accumulate sufficient quantity to qualify them as hyperaccumulators (Chapter 6). The accumulation of the heavy metals in the plants accounted for a rather small fraction (0.09 - 16%) of the overall heavy metal removal by the wetlands.

The *T. latifolia*-planted VSSF CW was used to remove total petroleum hydrocarbon (TPH), phenol, O&G, COD and TSS from secondary treated refinery wastewater (Chapter 7). The system achieved a removal efficiency from 45 to 100%. It was observed that the removal efficiency improved with age and maturity of the wetland plants and the CWs (Chapters 4, 5, and 6; Abira, 2008). This may be related to the increased aeration by extensive plant roots, adsorption and uptake (Abira, 2008) of the contaminants by *T. latifolia*. In addition, effective oil removal in CWs may be due to sorption onto soil particles and sedimentation. This may vary depending on temperature, plant biomass and the types and quantities of microorganisms present (Ji et al., 2007).

9.3.2. Macrophytes

The roles of wetland macrophytes and their functions as an essential component of the design of wetland systems are reviewed (Chapter 2). They have shown to be capable of removing pollutants from petroleum refinery wastewater (Mustapha et al., 2018; Mustapha et al., 2015; Chapter 5; Chapter 7 and Chapter 8). In CWs, a great diversity of microbial communities grow on wetland plant roots (Chapter 4) which enhances the pollutant removal efficiencies (Abou-Elela and Hellal, 2012). The choice of the type of macrophytes used in this study was based on the tolerance to high and low (fluctuating) loads of wastewater, plant productivity, rooting depth, availability and abundance in its locality. The major macrophytes found growing in and around the vicinity of KRPC were *C. alternifolius*, *C. dactylon* and *T. latifolia*. Their potential suitability for different types of wastewater treatment has been reported in many papers (Dhulap et al., 2014; Abou-Elela and Hellal, 2012; Kantawanichkukl et al., 2009; Calheiros et al., 2007), although only few studies are available for petroleum refinery wastewater treatment.

T. latifolia planted CWs had consistently higher removal efficiencies for all the measured parameters than *C. alternifolius* and *C. dactylon*-planted CW systems (Chapters 5 and 6). This may be due to the differences in the plant root penetration and larger root biomass hence larger root surface area (Fig. 9.2). Thus, based on the performance in the previous chapters (4, 5 and 6), *T. latifolia* was considered as a better choice and selected for further treatment of specific petroleum contaminants. However, in order to enhance the performance of the VSSF CW for the treatment of secondary treated refinery wastewater, 5 kg of cow dung was added on top of the substrate media as a slow releasing nutrient supplement to aid the phytoremediation processes. The resulting polished secondary treated refinery wastewater contained a lower concentration of TPH, O&G, phenol and COD, whereas BTEX was below the detection limits. Although, the manure improved the efficiency of the remediation processes, it increased the electrical conductivity, nitrogen and phosphorus content of the polished secondary treated refinery wastewater from the VSSF CWs (Chapter 7). Increased electrical conductivity values may be due to substrate - biofilm interactions, which may result in soluble salt release (Stefanakis and Tsihrintzis, 2012). In addition, high electrical conductivity values reduce the osmotic activity of plants and thus interfere with nutrients from the soil (Alobaidy et al., 2010; Camacho and Rodrigo, 2017). This explains the high concentrations of nitrogen and phosphorus in the effluent of the VSF CWs. The addition of cow dung to the *T. latifolia*-planted VSF CW stimulated plant biomass and petroleum compound tissue concentrations (Chapter 7). The increased *T. latifolia* biomass may have increased the mineralization of COD and the other measured contaminants (Zhu et al., 2015). The performance of the planted CW treatment systems was significantly higher than the unplanted control CWs for all the parameters measured (Chapters 4, 5, 6, 7 and 8). This suggests that the plants were able to oxygenate the treatment systems which lead to the aerobic degradation of organic loads in terms of the organic pollutants (Abdulhakeemet al., 2016). This finding has been reported by several researchers (Papaevangelou et al., 2017; Seeger et al., 2011; Taylor et al., 2011; Davis et al., 2009; Merkl et al., 2005; Tanner, 2001). The warm temperature in the tropics favours all year plant growth in the CWs (Katsenovich et al., 2009). Thus, the contribution of plant uptake to the overall nutrient removal is significant with subsequent harvesting of above ground biomass. The significance of plants was also evident from the TSS removal: *T. latifolia* and *C. alternifolius* had extensive roots (Fig. 9.2) that penetrated the entire wetland beds, thus they had a higher TSS removal efficiency compared to the *C. dactylon* CWs. Extensive roots provide a larger surface area, decrease water velocity enhancing sedimentation and filtration processes (Abdulhakeem et al., 2016; Brisson and Chazarenc, 2009). In addition, it was observed that the control (unplanted) CWs had a lower dissolved oxygen concentration than the planted CWs, an indication that plants contributed to oxygenation of the wetland beds.

The *T. latifolia*, *C. alternifolius* and *C. dactylon* biomass increased during the course of the experiment with negligible plant mortality. The plants showed a linear regression in both their growth and number of shoots (Chapter 5). *T. latifolia* showed the most rapid growth (R^2 = 0.9892), followed by *C. alternifolius* and then *C. dactylon*. *C. alternifolius* nevertheless had the highest number of shoots (R^2 = 0.9818), followed by *T. latifolia* (R^2 = 0.9686) and *C. dactylon* (R^2 = 0.9384). Thus, the micronutrients (Zn, Cu, Fe and Mn) as well as the essential nutrients (N and P) in the secondary treated refinery wastewater may have enhanced the growth of the plants in the CW.

9.3.3 Hybrid constructed wetlands

9.3.3.1. Hybrid constructed wetlands in Nigeria

A hybrid CW is a combination of two or more CWs in series that combines the advantages of the various CW systems to provide a better effluent quality than a single CW system (Zapater-Pereyra et al., 2015; Zurita and White, 2014; Ávila et al., 2013). In Chapter 5, a HSSF CW was coupled to the VSSF CW for further polishing of BOD_5, COD, ammonium-N, nitrate-N and phosphate-P removal from the secondary treated refinery wastewater. This gave an improved effluent quality, which can be attributed to the longer hydraulic retention time of the combined CWs (VSSF + HSSF), corresponding to an increased contact time for microbial degradation of organics (Otieno et al., 2017). The reduction of the concentrations of the contaminants mainly occurred in the VSSF CWs (first stage). The aerobic conditions present in the VSSF CWs may have contributed to BOD_5 (76%), COD (73%) and ammonium (70%) removal. In addition, the contribution of the HSSF CW to the overall removal processes was 9.8%, 18.8%, 27%, 13.9% and 32.2%, for respectively, BOD_5, COD, NH_4^+-N, NO_3^--N and PO_4^{3-}-P.

The hybrid CW showed a more improved effluent quality with removal efficiencies of 94% for BOD_5, 88% for COD, 84% for ammonium-N, 89% for nitrate-N, 78% for phosphate-P, 85% for TSS and 97% for turbidity. This improved effluent quality is due to the various advantages derived from a hybrid CW system. Many studies have reported low total phosphate removal in single CWs (Mustapha et al., 2015; Zurita and White, 2014; Chapter 5) as well as in hybrid systems (Zurita and White, 2014). However, this study showed a high phosphate removal (>78%) by the hybrid system (Chapter 5). This may be attributed to adsorption and precipitation processes, these are effective where wastewater continuously comes into contact with the filtration substrate (Zurita and White, 2014). The *T. latifolia*-planted hybrid CW had final effluent concentrations that conform very well to the discharge limits of the EU, WHO and FEPA (Nigeria): 5 NTU for turbidity, 30 mg/L for TSS, 10 mg/L for BOD_5, 40 mg/L for COD and 5 mg/L for phosphate. Thus, a hybrid CW is a viable alternative for the treatment of secondary refinery wastewater under the prevailing climatic conditions in Nigeria.

9.3.3.2. Duplex constructed wetlands

A duplex constructed wetland (duplex-CW) is a hybrid system that combines a vertical flow (VF) CW as a first stage and a horizontal flow filter (HFF) as a second stage for a more efficient wastewater treatment as compared to conventional constructed wetlands (Zapater-Pereyra et al., 2015). Diesel wastewater was treated in the duplex-CWs spiked with three different nutrient concentrations after 20 days of operation (Chapter 8). Also, the influent was supplied with artificial aeration to enhance the performance of the duplex-CWs (Chapter 8).

The VSSF compartments of the duplex-CW performed better than its HFF compartments (Chapter 8). Both *P. australis* and the artificial aeration may have improved the performance of the VSSF compartments of the duplex-CWs as compared to the performance of the HFF compartments (Chapter 8). The VSSF and HFF compartments were saturated with oxygen (Chapter 8; Table 8.2). The average influent dissolved oxygen entering the VSSF CW1 was 9.7 (± 0.5) mg/L and that entering the HFF1 was 4.6 (± 1.4) mg/L and the average dissolved oxygen in the effluent (outflow) was 3.0 (± 0.7) mg/L. Similarly, the average influent dissolved oxygen concentration entering the VSSF CW2 was 9.7 (± 0.5) mg/L and that entering the HFF2 was 4.6 (± 1.3) mg/L and the averaged dissolved oxygen in the effluent (outflow) was 3.2 (± 0.8) mg/L, while the VSSF CW3 had an average influent dissolved oxygen of 9.7 (± 0.5) mg/L, 4.0 (± 1.4) mg/L for the influent HFF3 and 2.8 (± 0.6) mg/L for the averaged outflow. Meanwhile, Zapater-Pereyra (2015) artificially aerated duplex-CW systems where the final dissolved oxygen of the VSF effluent was less than 2 mg/L. This was due to the air bubbles distribution, which decreased the chances of bed oxygenation that could have augmented the organic matter removal in the aerated systems.

The diesel wastewater contained low loads of COD (28.1 ± 28.1 to 31.7 ± 33.6 mg/L) and BTEX compounds (0.02 ± 0.01 to 0.22 ± 0.38 mg/L). The treated effluent COD concentration was below 3.6 (± 2.5) mg/L for VSSF CW1, 6.1 (± 5.9) mg/L for VSSF CW2 and 6.9 (± 5.6) mg/L for the VSSF CW3. Nutrient supplementation did not have an effect on the COD removal efficiency contrary to the literature (Fernandez-Luqueno et al., 2011; Eke and Scholz, 2008; Chapelle, 1999). The VSSF compartments (VF1, VF2 and VF3) showed a higher COD removal performance without macronutrient supplementation compared with their performance when nutrients were applied (Tables 8.4a and 8.4b). Thus, in this case, oxygenation has more effect on the COD removal efficiency than the nutrient addition. However, macronutrient supplementation enhanced the removal of petroleum hydrocarbons (BTEX) in the secondary treated refinery effluent (Table 8.4b; Chapter 8). This finding is in agreement with Eke and Scholz (2008) who also used *P. australis* in their experiment for benzene removal and concluded that addition of nutrients (10 - 60 mg/L NH_4^+ and 3 - 12 mg/L PO_4^{3-}) was more significant to benzene removal than the effects of plants or oxygen (Chapters 2 and 8).

The laboratory and environmental field duplex-CWs were highly effective for the treatment of hydrocarbons, nutrients and other water quality parameters. The systems were spiked with different concentrations of nutrients to enhance the treatment process. Numerical experiments were performed to investigate the biodegradation of the diesel compounds in the synthetic contaminated wastewater by the duplex-CWs using constructed wetland 2D (CW2D) (CW2D is a biokinetic model describing microbial dynamics as well as pollutant transformation and degradation processes in subsurface flow CWs (Pálfy et al., 2016)). The overall results showed a highly effective degradation of the diesel range organics. These results thus indicate that the CW2D model is a promising tool that can be used for real time monitoring and prediction of petroleum pollutant removal in duplex-CWs. The addition of nutrients to improve the performance of the diesel range organics removal efficiency in the duplex-CWs revealed a higher removal efficiency in the VF CWs compartment with no further treatment in the HFF compartments of the duplex-CW, while the days with no nutrient applications revealed a lower removal at the VF CWs compartment and additional removal at the HFFs compartment with a corresponding higher performance of the duplex-CWs (Chapter 8). This effective treatment of the diesel range organics thus makes the CW technology a viable and sustainable technology capable of meeting stringent water quality objectives (Chapter 8).

Fig. 9.2. The root system at the time of harvest of: (A) T. latifolia, (B) C. alternifolius and (C) C. dactylon growing in constructed wetlands treating secondary treated Kaduna refinery wastewater. Treatment of performance of these constructed wetlands is described in chapters 4, 5, 6 and 7.

9.4. Implications for application by petroleum refining industry

The characterized secondary treated refinery wastewater contained pollutants that were above the permissible discharge limits (Chapter 3) allowable by the regulating authorities (FEPA, EU and WHO). The effect of discharging partially or poorly treated petroleum effluents into the environment leads to pollution of the terrestrial and aquatic ecosystems, deteriorating water quality which can lead to health problems and diseases due to increased levels of hazardous constituents into the water bodies (Chapter 4; Wake, 2005). For instance, TDS contain nutrients, toxic metals and organic pollutants (Murray-Gulde et al., 2003), thus discharging effluent into the receiving stream with a high TDS content can cause toxicity through the increase in salinity, changes in the ionic composition of the water, and the toxicity of individual compounds while BOD_5, COD, ammonium and phosphate cause oxygen depletion (Abira, 2008). Heavy metals in discharged wastewater are hazardous to public health (Mustapha et al., 2018; Yadav et al., 2012; Lesage, 2007) and high phenol concentrations in discharged effluents are toxic to fishes (Abira, 2008). Although the secondary treated refinery wastewater contains medium strength nutrients, due to the large volume of wastewater discharge, there is a potential for eutrophication of the Romi River, including the River Kaduna downstream.

Other effects of the discharge of partially treated wastewater includes an unaesthetic environment, it can lead to high costs of wastewater treatment, loss of farmlands, fishes, potable water, and means of the livelihood of the affected community in particular. Thus, removal of BOD$_5$, COD, nutrients, heavy metals, phenol, O&G, TPH from the discharged secondary treated refinery wastewater is very important. However, conventional physicochemical approaches are expensive especially at large scale (Yadav et al., 2012; Chapter 4). In addition, the conventional methods used by KRPC (Nigeria) cannot sufficiently remove the contaminants of concerns to compliance limits (FEPA, WHO and EU). In order to improve the wastewater quality, meet stringent guidelines as well as protect the recipient streams (Romi River and downstream rivers), installation of CWs at the effluent discharge point of the KRPC (Nigeria) is strongly recommended.

9.5. Scaling up of constructed wetland for treating petroleum refinery effluent

9.5.1. Design parameters

The water budget and influent characterization are required for the design of a full scale constructed wetland (ITRC, 2003). It is a quantification of all the water flowing into and out of the wetlands. The inflow includes natural flows, process flows, storm water runoff, precipitation, ice thaws and groundwater. The outflows are evaporation, transpiration, wetland outlet system and possibly groundwater (ITRC, 2003). The water budget is calculated based on the equation as given in Leto et al. (2013):

$$Q_o = Q_i + (P - ET) A \text{ --- (9.1)}$$

where Q_o is the output wastewater flowrate (m^3/d), Q_i is the wastewater inflow rate (m^3/d), P is the precipitation rate (m/d), ET is the evapotranspiration rate (m/d) and A is the wetland top surface area (m^2).

The performance of a tropical treatment wetlands is better evaluated considering the effects of rainfall and evapotranspiration on mass pollutant load removal (Chapter 5; Dhulap et al., 2014; Katsenovich et al., 2009). The average inflow and outflow were given as 98,856 m^3/day and 85,575 m^3/day, respectively, based on equation (9.1).

Influent characterization

This entails the quantity and quality of the water to be treated. The quantity of the wastewater has been determined (Table 9.2). KRPC (Kaduna) generates about 100, 000 m^3 of wastewater per day from different sources which include process operations, cooling tower blowdown, tank drainage and storm water runoff. The breakdown of the wastewater volume generated is presented in Table 9.2. From Chapter 3 of this thesis, the influent has been characterized, the concentrations of the parameters of concern are presented in Table 2.2. The treatment objective is to meet discharge standards (FEPA (Nigeria), EU and WHO).

Influent concentration, mg/L

BOD_5 is chosen as the contaminant of interest for the design of the full scale CW, based on its concentrations (mean highest value). BOD_5 = 107 mg/L based on the characterization result collected during a period of 16 months from September 2011 to December 2012 (Chapter 3).

The target effluent concentration, mg/L

The effluent discharge target is chosen as 10 mg/L based on Federal Environmental Protection Agency (FEPA), Nigeria permissible discharge limit and on KRPC effluent discharge target.

Hydraulics

Hydraulic retention time, HRT, days:

HRT = (Wetland volume, m^3) / (Average flow rate, m^3/day) ------------------ (9.2)

Hydraulic loading rate, HLR m^3/m^2 day or m/day

HLR = (Average flow rate, m^3/day) / (wetland area, m^2) ----------------------- (9.3)

The areal loading rates (ALR)

The areal loading rates (ALR) can also be used for sizing the wetland for both planning and final design of the wetland systems from the pollutant mass loads and effluent requirements (Chapters 3, 4, 5, 6). The ALR is the maximum removal rate of a pollutant mass per unit surface area of a wetland per daily input (ITRC, 2003):

ALR = (Influent flowrate, m³/day) * (pollutant concentration, mg/L) / (surface area, m² of the subsurface flow wetland) -- (9.4)

Depth, m

Based on the results of the length of the roots of *T. latifolia* and *C. alternifolius* in this study, and for effective or complete removal of TSS and BOD, the media depth of 0.8 – 1.0 m and water depth of 0.6 – 0.8 m are recommended for the full scale CW.

Site suitability

Constructed wetlands require a large area. KRPC (Kaduna) has a large area for siting a full scale constructed wetlands. Kaduna is located in the tropics, which supports plant growth all year round (Katsenovich et al., 2009). The average water temperature is 26 ^{0}C (Table 3.2, Chapter 3)

Size of full scale CWs

CW. The effective required treatment area can be calculated based on preliminary information. The average flow rate or discharge effluent per day = 98, 856 m³ (Table 9.2).

The first-order plug-flow kinetics can be used to estimate BOD_5 removal in SSF CWs (US EPA, 1988). This equation can be re-arranged to estimate the required surface area for a subsurface flow CW system.

$$[C_e/C_o] = \exp(-K_T t) \text{ --9.5}$$

$$As = \frac{[Q(\ln C_o - \ln C_e)]}{k_T \times d \times n} \text{ -- 9.6}$$

where As = surface area of the system, m²

      Q = average flow rate through the system, m³/d

      Co = influent BOD_5, mg/L

Ce = effluent BOD₅, mg/L

K_T = temperature-dependent first-order reaction rate constant, d^{-1}

d = depth of submergence, m

n = porosity of the bed, as a fraction

t = hydraulic residence time, d

The design parameters are as follows:

1. *T. latifolia* is chosen for this design consideration since it grew better than *C. alternifolius* and *C. dactylon*. From chapter 7, the root depth of *T. latifolia* was measured as 0.45 m at harvest.

2. A bed slope of 1% is selected for ease of construction (s = 0.01).

3. k_s S < 8.60 (US EPA, 1988). Thus, coarse sand, n = 0.39, k_s = 480 and K_{20} = 1.35 is chosen for this design (k_s = hydraulic conductivity of the medium, $m^3/m^2.d$; S = slope of the bed or hydraulic gradient (as a fraction or decimal)):

$$k_s \text{ S} \text{ ---9.7}$$
$$k_s \text{ S} < 8.60 = (480)(0.01) = 4.8 < 8.60$$

4. First-order temperature-dependent rate constant (K_T) using the equation:

$$K_T = K_{20} (1.1)^{T-20} \text{ --9.8}$$

where K_{20} is the rate constant @ 20°C

From preliminary studies, Chapter 3, temperature ranges 22 – 31°C.

At minimum temperature:

$$K_T = 1.35 (1.1)^{22-20} = 1.63$$

At maximum temperature:

$$K_T = 1.35 (1.1)^{31-20} = 3.85$$

5. The cross sectional area (Ac) of the bed is given as:

$$Ac = \frac{Q}{k_s S} \text{ ---9.9}$$
$$Ac = \frac{98,856}{(480)(0.01)} = 20,595 \text{ m}^2$$

6. The bed width (W) is given as:

$$W = \frac{A_c}{d} \text{ ---9.10}$$
$$W = \frac{20,595}{0.45} = 45,767 \text{ m}$$

7. The surface area (As) required can be calculated from equation 9.6. From equation 9.6, at minimum and maximum temperature, K_T = 1.68 and 3.85, respectively.

K_T = 1.68

$$As = \frac{[\,98,856\,(\,In\,107 - In\,10)]}{1.63 \times 0.45 \times 0.39}$$

$$As = \frac{[\,98,856\,(\,4.67 - 2.30)]}{1.63 \times 0.45 \times 0.39} = \frac{234,288.72}{0.286} = 819,191.33\ m^2 = 81.92 = 82\ ha$$

K_T = 3.85

$$As = \frac{[\,98,856\,(\,In\,107 - In\,10)]}{3.85 \times 0.45 \times 0.39}$$

$$As = \frac{[\,98,856\,(\,4.67 - 2.30)]}{3.85 \times 0.45 \times 0.39} = \frac{234,288.72}{0.676} = 346,580.95\ m^2 = 34.66 = 35\ ha$$

For this design, we will choose the lower temperature corresponding to the larger surface area (82 ha).

8. The bed length (L) and the detention time (t) is given as:

$$L = \frac{A_s}{W} \quad \text{--9.11}$$

$$L = \frac{819,191.33}{45,767} = 17.90\ m$$

$$t = \frac{V_v}{Q} = \frac{LWdn}{Q} \quad \text{--9.12}$$

$$t = \frac{(17.90)(45,767)(0.45)(0.39)}{98,856} = \frac{143,774.74}{98,856} = 1.45\ days$$

9. The required width will be divided into individual cells of 915 m wide for better hydraulic control at the inlet zone. 50 cells will be constructed, each 915 m x 18 m.

Types of CW

Both VSSF and HSSF CWs were effective in the treatment of secondary treated refinery wastewater. Also, SSF CWs will reduce smell and odour and proliferation of the nuisance insects such as mosquitoes and flies. Therefore, both of these wetlands are recommended for use. However, due to the calculated treatment area, the CWs should be connected in series for easy assessment and maintenance.

Macrophytes

T. latifolia, C. alternifolius and *C. dactylon* are locally available plant species. These plants were found growing in and around the vicinity of KRPC (Kaduna). The adaptation of *T. latifolia, C. alternifolius* and *C. dactylon* for the full scale CWs is recommended, since the three plants were effective in the treatment of secondary treated refinery wastewater. In addition, they have a fast growth rate, good rooting depth and are perennial plants (Chapter 5). Also, the use of multiple culture plants is recommended since the different plants play different roles in the purification of pollutants.

Substrate media

The treatment media used in this study was gravel mixed with coarse sand. It has a porosity of 40 % (Chapter 4). This produced void ratios with sufficient retention time that achieved the target removal rates of contaminants (Chapter 4, 5, 6, 7 and 8). Therefore, gravel mixed with coarse sand is recommended for the full scale constructed wetlands.

9.5.2. Operation and maintenance

Flow mode

Continuous flow mode is recommended for the full scale CW for the treatment of secondary treated refinery wastewater discharged by KRPC (Nigeria). Although, only continuous flow mode was used in this study, Abira (2008) recommended a continuous flow mode for full scale wastewater treatment since their wetland had a lower removal efficiency and poorer plant growth when batch operation mode was used.

Maintenance

It is expected that continuous maintenance is necessary to sustain a high purification efficiency as well as prolong the life of the wetland systems (Abira, 2008). The potential maintenance activities include inlet and outlet structures to ensure uniformity of water, control of the water level (determines the functioning of wetland), management of vegetation and maintenance of water control structures.

Table 9.3. Summary of design criteria for subsurface flow constructed wetlands for polishing secondary treated refinery wastewater

Parameter	Criteria	Remarks
Water budget		Calculated (Equation 9.1)
Q_i	98,856 m³/day	
Q_O	85,575 m³/day	
Influent concentrations	107 mg BOD/L	This study (Table 3.2 in Chapter 3)
Effluent concentrations	< 10 mg/L	Based on the performance of this study (Chapters 4 and 5) and the regulatory authorities (FEPA, Nigeria, KRPC limits)
Surface area	Based on average influent BOD concentrations	This study (Table 3.2 in Chapter 3)
Required area	82 ha	Calculated (Equation 9.6)
Root depth:	0.45 m	This study (Chapter 7)
Media	0.8 - 1.0 m	This study (Chapter 8)
Water	0.6 – 0.8 m	
Media: Gravel mixed with coarse sand	Media arrangement: Top layer (inlet): 6 – 10 mm Middle layer: 16 – 25 mm Bottom layer (outlet): 25 – 36 mm	This study (Chapters 4; 5 and 6)
Constructed wetland train	Use 50 SSF CWs in parallel and in series	This study (hybrid system) (Chapter 5)
Flow mode	Continuous	
Macrophytes	*T. latifolia*, *C. alternifolius* and *C. dactylon*	This study (Chapters 4, 5, 6 and 7)
Miscellaneous	Use adjustable inlet and outlet device to balance flow	This study (Chapter 4)

(Source: Modified from ITRC, 2003 and Abira, 2008).

Monitoring

To ensure proper and continuous functioning of the wetlands, regular monitoring of the wetlands is recommended. The necessary measures will be put in place to ensure that the wetland treatment system is performing well and that the discharged effluent meets the compliance limits. Weekly (pH, DO, temperature and electrical conductivity), bimonthly (BOD, COD, TSS), monthly (TPH, BTEX, phenol, oil and grease and selected heavy metals) and yearly (plant cover) monitoring of the different parameters of concern is recommended for the full scale constructed wetland.

Harvested macrophytes

The localities/local farmers around the wetlands can manage and harvest the macrophytes. The harvested macrophytes can be put to other uses such as raw material for industrial products including paper mill, roof tiles, plant biomass fuel as supplementary energy source, stuffing quilt and dolls. The investigation of multi-culture macrophytes for polishing secondary refinery effluents or combinations of *T. latifolia, C. alternifolius* and *C. dactylon* (e.g. *T. latifolia* + *C. alternifolius, T. latifolia* + *C. dactylon*, C. *alternifolius* + *C. dactylon*) in the effective removal of pollutants from petroleum refinery wastewater should be investigated, since different plants have different abilities (Tam and Wong, 2014; Stottmeister et al., 2003). The combined effects of a poly-cultured constructed wetland might improve its treatability. More so, there are few studies on poly-cultured constructed wetlands (Tam and Wong, 2014), it therefore makes it a good research opportunity.

The disposal of the biomass of *T. latifolia, C. alternifolius* and *C. dactylon* after wastewater treatment is a critical research area to consider as well. The harvested biomass from CWs can provide economic returns to communities for biogas production, biofertilizer, biomaterial and animal food (Li et al., 2007) as well as fibre for paper making and compost (Belmont et al., 2004). Harvested biomass is a good source of plant biomass fuel, according to Pratt et al. (1988), the energy content of the aboveground *T. latifolia* biomass is between 17.6 and 18.9 MJ/kg and in Indiana (USA), Apfelbaum (1985) reported that *T. latifolia* contributed 700 kg of biomass per hectare where it grew in monocultures. These economic benefits from CWs are an important consideration in developing countries where additional incentives are required to encourage the communities to maintain treatment wetlands.

9.5.3. Cost implication for siting a subsurface flow constructed wetland to polish the secondary treated Kaduna refinery wastewater

Constructed wetlands have low construction costs compared to conventional wastewater treatment technologies (Chapter 2; Abira, 2008). The main investment costs to consider for the construction of wetlands are land requirement, site survey, system design, site preparations, plastic liners for prevention of ground water contamination, filtration, rooting media, vegetation, hydraulic control structures and miscellaneous costs which may include fencing and access roads (Chapter 2). Table 9.4 present a rough cost estimate for a 4645 m² subsurface flow constructed wetland.

Land availability is not a constraint for KRPC (Nigeria). The cost of gravel material is relatively high. The gravel cost is about 40 - 50% of the system for a 4645 m² system with the percentage increasing as the system gets larger (ITRC, 2003). The liner is about 15 - 25% of the total costs, with this percentage decreasing as the system gets larger (ITRC, 2003). Excavation/earthwork is estimated as the third or fourth largest cost of the project cost, with flat areas less costly. Plants are a minor cost as aquatic emergent macrophytes are generally available (Chapters 4 and 5). Alternatively, plants can be bought from nurseries that are capable of providing the quantity, species and quality of plants needed. Planting in gravel is easy. Other costs include piping, level control structures, flow distribution structures, flow meters and fencing.

Table 9.4. Cost of a 4645 m² subsurface flow constructed wetland

Component	Price/unit ($)*	Total ($)	%
Excavation/compaction	1.75 /m³	13,000	11
Gravel	16 /m³	51,900	44
Liner	35 cent/m²	19,250	16
Plants	60 cent/each	13,300	11
Plumbing		7,500	07
Control structures		7,000	06
Miscellaneous		10,000	05
Total		121,980	

*(Source: Modified from ITRC, 2003).

9.6. Outlook: Application of constructed wetland treatment technology to KRPC effluent

The main objective of this PhD study was to polish the secondary refinery wastewater discharged by KRPC (Nigeria) into the environment to below non-hazardous (compliance) limits in order to improve the quality of the receiving water. This goal was achieved by the sets of the designed experiments and this technology can be adopted by KRPC (Nigeria) to meet its stringent water quality objectives.

The CW treatment technology is simple, cost effective, can easily be maintained when compared with conventional methods. The technology can be adopted for use by KRPC (Kaduna, Nigeria) and other developing nations where effective, low cost wastewater treatment strategies are critically needed. Therefore, the application of CWs for polishing KRPC wastewater will assist in reducing the level of hazardous constituents into water bodies as well as soil and assure improved water quality by the discharge of treated wastewater into the environment. The adequately treated wastewater from CW systems can be reused and/or safely discharged into water bodies, this can drastically reduce the cost of potable water production. In addition, fewer polluted water bodies will ensure an environment that is conducive for fishes and all forms of aquatic life as well as improve the livelihood of the affected community. Hence, health problems and diseases associated with the discharge of untreated or inadequately treated wastewater will be minimized.

9.7 References

Abdulhakeem, S. G., Aboulroos, S. A., Kamel, M. M. (2016). Performance of a vertical subsurface flow constructed wetland under different operational conditions. Journal of Advanced Research, 7, 803-814. Retrieved from http://dx.doi.org/10.1016/j.jare.2015.12.002

Abira, M. A. (2008). A pilot constructed treatment wetland for pulp and paper mill wastewater: performance, processes and implications for the Nzoia River, Kenya. Delft, The Netherlands: The Academic Board of the Wageningen University and the UNESCO-IHE, Delft.

Abou-Elela, S. I., Hellal, M. S. (2012). Municipal wastewater treatment using vertical flow constructed wetlands planted with *Canna*, *Phragmites* and *Cyprus*. Ecological

Engineering, 47, 209-213. Retrieved from
http://dx.doi.org/10.1016/j.ecoleng.2012.06.044

Alobaidy, A. H., Al-Sameraiy, M. A., Kadhem, A. J., Abdul Majeed, A. (2010). Evaluation
of treated municipal wastewater quality for irrigation. Journal of Environmental
Protection, 1, 216-225. doi:10.4236/jep.2010.13026

Apfelbaum, S. I. (1985). Cattail (*Typha* spp.) management. Natural Area Journal, 5(3), 9-17.

Aslam, M. M., Malik, M., Braig, M. A. (2010). Removal of metals from refinery wastewater
through vertical flow constructed wetlands. International Journal of Agriculture and
Biology, 12, 796-798.

Ávila, C., Garfí, M., García, J. (2013). Three-stage hybrid constructed wetland system for
wastewater treatment and reuse in warm climate regions. Ecological Engineering, 43-
49. doi:10.1016/j.ecoleng.2013.09.048

Belmont, M. A., Cantellano, E., Thompson, S., Williamson, M., S´anchez, A. M. (2004).
Treatment of domestic wastewater in a pilot-scale natural treatment system in central
Mexico. Ecological Engineering, 23, 299–311. (doi:10.1016/j.ecoleng.2004.11.003)

Brisson, J., Chazarenc, F. (2009). Maximizing pollutant removal in constructed wetlands:
Should we pay more attention to macrophyte species selection? Science of The Total
Environment, 407(13), 3923-3930. doi:10.1016/j.scitotenv.2008.05.047

Calheiros, C., Rangel, A., Castro, P. (2007). Constructed wetland systems vegetated with
different plants applied to the treatment of tannery wastewater. Water Res, 41, 1790 -
1798. doi:10.1016/j.watres.2007.01.012

Camacho, V. J., Rodrigo, M. A. (2017). The salinity effects on the performance of
constructed wetland-microbial fuel cell. Ecological Engineering, 107, 1-7. Retrieved
from https://doi.org/10.1016/j.ecoleng.2017.06.056

Chapelle, F. (1999). Bioremediation of petroleum hydrocarbon-contaminated ground water:
The perspectives of history and hydrology. Ground Water , 37(1), 122-132.
doi:10.1111/j.1745-6584.1999.tb00965.x

Davis, B. M., Wallace, S., Willison, R. (2009). Pilot-scale engineered wetlands for produced water Treatment. Society of Petroluem Engineers, 4(3), 75-79. Retrieved December 22, 2016, from http://dx.doi.org/10.2118/120257-PA

Dhulap, V. P., Ghorade, I. B., Patil, S. (2014). Seasonal study and its impact on sewage treatment in the angular horizontal subsurface flow constructed wetland using aquatic macrophytes. International Journal of Research in Engineering & Technology, 2(5), 213-224.

Dipu, S., Thanga, S. G. (2009). Bioremediation of industrial effluents using constructed wetland technology. Retrieved February 1, 2011, from http://www.eco-web.com/edi/090227.html

EU. (2014). Eupean Union (Drinking Water) Regulations 2014. EU. Retrieved January 9, 2017, from https://www.fsai.ie/uploadedFiles/Legislation/Food_Legisation_Links/Water/SI122_2014.pdf

Eze, P. E., Scholz, M. (2008). Benzene removal with vertical - flow constructed treatment wetlands. Journal of Chemical Technology and Biotechnology, 83(1), 55-63. doi:10.1002/jctb.1778

FEPA. (1991). Guidelines and standards for environmental pollution control in Nigeria. National Environmental Standards-Parts 2 and 3. Lagos, Nigeria: Government Press.

Fernandez-Luqueno, F., Valenzuela-Encinas, C., Marsch, R., Martinez-Suarez, C., Vazquez-Nunez, E., Dendoven, L. (2011). Microbial communities to mitigate contamination of PAHs in soil-possibilities and challenges: a review. Environ. Sci. Pollut Res. , 12-30. doi:10.1007/s11356-010-0371-6

Horner, J. E., Castle, J. W., Rodgers Jr., J. H., Murray Gulde, C., Myers, J. E. (2012). Design and performance of pilot-scale constructed wetland treatment systems for treating oilfield produced water from sub-saharan africa. Water Air Soil Pollut, 223, 1945-1957. doi:10.1007/s11270-011-0996-1

ITRC (Interstate Technology and Regulatory Council) (2003). Technical and regulatory guidance document for constructed treatment wetlands. Interstate Technology and Regulatory Council. Retrieved from http://www.itrcweb.org.

Ji, G., Sun, T., Ni, J. (2007). Surface flow constructed wetland for heavy oil-produced water treatment. Bioresource Technology, 98(2), 436-441. doi:10.1016/j.biortech.2006.01.017

Kantawanichkukl, S., Kladprasert, S., Brix, H. (2009). Treatment of high-strength wastewater in tropical vertical flow constructed wetlands planted with *Typha angustifolia* and *Cyperus involucratus*. Ecological Engineering, 35, 238-247. doi:10.1016/j.ecoleng.2008.06.002

Katsenovich, Y. P., Hummel-Batista, A., Ravinet, A. J., Miller, J. F. (2009). Performance evaluation of constructed wetlands in a tropical region. Ecological Engineering, 35(10), 1529-1537. doi:10.1016/j.ecoleng.2009.07.003

Kivaisi, A. K. (2001). The potential for constructed wetlands for wastewater treatment and reuse in developing countries: a review. Ecol. Eng., 16, 545-560 . doi:PII: S0925-854(00)00113-0

Lesage, E., Rousseau, D., Meers, E., Tack, F., De Pauw, N. (2007). Accumulation of metals in a horizontal subsurface flow constructed wetland treating domestic wastewater in Flanders, Belgium. Science of the Total Environment, 380, 102-115. doi:10.1016/j.scitotenv.2006.10.055

Leto, C., Tuttolomondo, T., La Bella, S. L. (2013). Effects of plant species in a horizontal subsurface flow constructed wetland –phytoremediation of treated urban wastewater with *Cyperus alternifolius* L. and *Typha latifolia* L. in the West of Sicily (Italy). Ecological Engineering, 61, 282-291. doi:10.1016/j.ecoleng.2013.09.014

Li, M., Wu, Y.-J., Yu, Z.-L., Sheng, G.-P., Yu, H.-Q. (2007). Nitrogen removal from eutrophic water by floating-bed-grown water spinach (*Ipomoea aquatica Forsk.*) with ion implantation. Water Research, 41, 3152-3158. doi:10.1016/j.watres.2007.04.010

Liu, J., Dong, Y., Xu, H., Wang, D., Xu, J. (2007). Accumulation of Cd, Pb and Zn by 19 wetland plant species in constructed wetland. Hazard. Mat., 147(3), 947-953. doi::10.1016/j.jhazmat.2007.01.125

Madera-Parra, C. A., Pena-Salamanca, E. J., Pena, M., Rousseau, D. P., Lens, P. N. (2015). Phytoremediation of landfill leachate with *Colocasia esculenta, Gynerum sagittatum* and *Heliconia psittacorum* in constructed wetlands. International Journal of Phytoremediation, 17, 16-24. doi:10.1080/15226514.2013.828014

Merkl, N., Schultze-Kraft, R., Infante, C. (2005). Assessment of tropical grasses and legumes for Phytoremediation of petroleum-contaminated soils. Water Air Soil Pol., 165, 195-209.

Murray-Gulde, C., Heatley, J. E., Karanfil, T., Rodgers Jr, J. R., Myers, J. E. (2003). Performance of a hybrid reverse osmosis-constructed wetland treatment system for brackish oil field produced water. Water Research, 37(3), 705-713. PII:S0043-1354(02)00353-6

Mustapha, H. I., Rousseau, D., van Bruggen, J., Lens, P. (2011). Treatment performance of horizontal subsurface flow constructed wetlands treating inorganic pollutants in simulated refinery effluent. In: 2nd Biennial Engineering Conference. School of Engineering and Engineering Technology, Federal University of Technology, Minna.

Mustapha, H. I., van Bruggen, J. J., Lens, P. N. (2015). Vertical subsurface flow constructed wetlands for polishing secondary Kaduna refinery wastewater in Nigeria. Ecological Engineering, 84, 588-595. http://dx.doi.org/10.1016/j.ecoleng.2015.09.060.

Mustapha, H. I., van Bruggen, J. J., Lens, P. N. (2018). Fate of heavy metals in vertical subsurface flow constructed wetlands treating secondary treated petroleum refinery wastewater in Kaduna, Nigeria. International Journal of Phytoremediation, 20(1), 44-53. doi:10.1080/15226514.2017.1337062

Nouri, J., Khorasani, N., Lorestani, B., Karami, M., Hassani, A. H., Yousefi, N. (2009). Accumulation of heavy metals in soil and uptake by plant species with phytoremediation potential. Environ Earth Science, 59(2), 315-323. doi:10.1007/s12665-009-0028-2

Otieno, A. O., Karuku, G. N., Raude, J. M., Koech, O. (2017). Effectiveness of the horizontal, vertical and hybrid subsurface flow constructed wetland systems in polishing municipal wastewater. Environmental Management and Sustainable Development, 6(2), 2164-7682. doi:10.5296/emsd.v6i2.11486

Pálfy, T. G., Molle, P., Langergraber, G., Troesch, S., Gourdon, R., Meyer, D. (2016). Simulation of constructed wetlands treating combined sewer overflow using

HYDRUS/CW2D. Ecological Engineering, 87, 340-347. Retrieved from http://doi.org/10.1016/j.ecoleng.2015.11.048

Papaevangelou, V. A., Gikas, G. D., Tsihrintzis, V. A. (2017). Chromium removal from wastewater using HSF and VF pilot-scale constructed wetlands: Overall performance, and fate and distribution of this element within the removal environment . Chemosphere, 168, 716-730. http://dx.doi.org/10.1016/j.chemosphre.2016.11.002

Pratt, D. C., Dubbe, D. R., Garver, E. G., Johnson, W. D. (1988). Cattail (*Typha* spp.) biomass production - stand management and sustainable yields. University of Minnesota, Bio-Energy Coordinating Office and Department of Botany. Crookston: Oak Ridge National Laboratory. Retrieved June 3, 2016

Schroder, P., Navarro-Avino, J., Azaeizeh, H., Goldhirsh, A. G., DiGregorio, S., Komives, T., Langergraber, G., Lenz, A., Maestri, E., Memon, A. R., Ranalli, A., Sebastiani, L., Smreck, S., Vanek, T., Vuilleumier, S., Wissing, F. (2007). Using Phytoremediaion Technologies to Upgrade Wastewater Treatment in Europe. Environmental Science and Pollution Research-International, 14(7), 490-497. doi:10.1065/espr2006.12.373

Seeger, E. A., Kuschk, P., Fazekas, H., Grathwohl, P., Kaestner, M. (2011). Bioremediation of benzene-, MTBE- and ammonia-contaminated groundwater with pilot-scale constructed wetlands. Environmental Pollution, 159, 3769-3776. doi::10.1016/j.envpol.2011.07.019

Shpiner, R., Vathi, S., Stuckey, D. C. (2009). Treatment of "produced water" by waste stabilization ponds: removal of heavy metals. Water Research, 43, 4258-4268. doi:10.1016/j.watres.2009.06.004

Simi, A. (2000). Water quality assessmentof a surface flow constructed wetland treating oil refinery wastewater. In K. R. Reddy (Ed.), 7th International Conference on Wetlands System for Water Pollution Control. 3, pp. 1295-1304. Lake Buena Vista, Boca Raton, Florida, USA: IWA. Retrieved January 17, 2011

Stefanakis, A. I., Tsihrintzis, V. A. (2012). Effects of loading, resting period, temperature, porous media, vegetation and aeration on performance of pilot-scale vertical flow constructed wetlands. Chemical Engineering Journal, 181-182, 416-430. doi:10.1016/j.cej.2011.11.108

Stottmeister, U., Wießner, A., Kuschk, P., Kappelmeyer, U., Kästner, M., Bederski, O., Müller, R. A., Moormann, H. (2003). Effects of plants and microorganisms in constructed wetlands for wastewater treatment. *Biotechnology Advances, 22*, 93-117. doi:10.1016/j.biotechadv.2003.08.010

Tam, N. F.-Y., Wong, Y.-S. (2014). Constructed wetland with mixed mangrove and non-mangrove plants for municipal sewage treatment. 2014 4th International Conference on Future Environment and Energy. 61, pp. 60-63. Singapore: IACSIT Press. doi:10.7763/IPCBEE

Tanner, C. C. (2001). Plants as ecosystem engineers in subsurface-flow treatment wetlands. Water Science and Tehnology, 44(11-12), 9-17.

Taylor, R. C., Hook, B. P., Stein, R. O., Zabinski, A. C. (2011). Seasonal effects of 19 plant species on COD removal in subsurface treatment wetland microcosms. Ecol. Eng., 37(5), 703-710. doi:10.1016/j.ecoleng.2010.05.007

US EPA (Environmental Protection Agency). (1988). Design manual - constructed wetlands and aquatic plant systems for municipal wastewater treatment. USA: U.S. Environmental Protection Agency, Office of Research and Development, Center for Environmental Research Information, Cincinnati, OH 45268. doi:EPA/625/1-88/022

Vymazal, J. (2011). Constructed wetlands for wastewater treatment: five decades of experience. Environ. Sci. Technol., 45(1), 61-69. doi:10.1021/es101403q

Vymazal, J. (2011). Plants used in constructed wetlands with horizontal subsurface flow: a review. Hydrobiologia, 674, 133-156. doi:10.1007/s10750-011-0738-9

Wake, H. (2005). Oil refineries: a review of their ecological impacts on the aquatic environment. Estuarine Coastal and Shelf Science, 62, 131-140. doi:10.1016/j.ecss.2004.08.013

Wallace, S., Schmidt, M., Larson, E. (2011). Long Term Hydrocarbon Removal Using Treatment Wetlands. SPE Annual Technical Conference and Exhibition (pp. 1-10). Denver, Colorado: Society of Petroleum Engineers (SPE) International.

WHO (2008). Guidelines for drinking water quality (3rd ed.). Geneva: Health Criteria and Supporting Information.

Yadav, A. K., Abbassi, R., Kumar, N., Satya, S., Sreekrishnan, T. (2012). The removal of heavy metals in wetland microcosms: Effects of bed, plant species, and metal mobility. Chemical Engineering Journal, 211-212, 501-507. doi:10.1016/j.cej.2012.09.039

Zapater-Pereyra, M. (2015). Design and development of two novel constructed welands: the duplex-constructed wetland and the constructed wetroof. Leiden, The Netherlands: CRC Press/Balkema.

Zapater-Pereyra, M., Ilyas, H., S, L., van Bruggen, J., Lens, P. (2015). Evaluation of the performance and the space requirement by three different hybrid constructed wetlands in a stack arrangement. Ecological Engineering, 82, 290-300.

Zhu, B., Panke-Buisse, K., Kao-Kniffin, J. (2015). Nitrogen fertilization has minimal influence on rhizosphere effects of smooth crabgrass (*Digitaria ischaemum*) and bermuda grass (*Cynodon dactylon*). Journal of Plant Ecology, 8(4), 390-400. doi:10.1093/jpe/rtu034

Zurita, F., White, J. R. (2014). Comparative study of three two-stage hybrid ecological wastewater systems for producing high nutrient, reclaimed water for irrigationreuse in developing countries. *Water, 6*, 213-228. doi:10.3390/w6020213

Appendix

APPENDIX A

Mass balance approach

This approach was used to identify the main removal pathways of heavy metals in vertical subsurface flow (VSSF) CWs fed with secondary treated refinery wastewater. The number of plants, shoots and height of *Typha latifolia*, *Cyperus alternifolius* and *Cynondon dactylon* were observed and monitored during the experimental period. At the start (day 0), day 90, day 180 and day 270 of the experiment, the three plants biomass were harvested and divided into roots, stems and leaves. These plant parts were dried to constant weight at 105^0C for heavy metal analysis (Section 6.2.3). Wastewater samples were collected every fourteen days from both the inlet and outlet of the VSSF CWs for heavy metal analysis (Section 6.2.2).

Table A1 present the mass balance of heavy metals in VSSF CWs fed with secondary treated refinery wastewater. There was a decrease of heavy metal concentrations in the effluents from all the three planted CWs including the unplanted CWs. Meanwhile, there was increase in heavy metal concentrations in the plants parts as well as in the sediments (Table A1). The planted CWs had higher heavy metal removal efficiency compared with the unplanted CWs, with higher concentrations of heavy metals in their sediments.

The removal efficiency was calculated as the percentage of removal for each heavy metal as:

$$\text{Removal efficiency} = \frac{CiVi - CeVe}{CiVi} \times 100$$

Where C_i and C_e are the influent and effluent concentrations in mg/L and V_i and V_e are the volume of the influent and effluent in litres fed and collected in the VSSF CWs. The contributions (Tables A1 and A2) made by the plants, sediment and other sources are calculated as:

$$\text{Removal (plant, sediment)} = \frac{\text{Mass, mg}}{\text{Total output, mg}} \times 100$$

Less than 5% of the heavy metals accumulated in the three plants were mainly retained in the belowground biomass. Although the daily input of heavy metals in the influent were low (Tables 3.3 and 6.1) the sediment was shown as an important sink for heavy metals in the VSSF CWs fed with secondary treated refinery wastewater (Table A2).

Table A1. Mass balance of heavy metals in vertical subsurface flow constructed wetlands

Plant	Heavy metal	Influent, mg	Root (initial), mg	Stem (initial), mg	Leaf (initial)s, mg	Sediment (initial), mg	Effluent, mg	Root (final), mg	Stem (final), mg	Leaf (final), mg	Sediment ((final), mg	Total input, mg	Total output, mg	Removal,%
T.latifo	Cr	82683	9.43	0.15	0.32	0.0	9401	147.58	17.40	10.20	6392	82693	15968	80.7
C.	Cr		15.93	1.81	1.16	0.0	16851	79.30	12.73	4.11	5532	82701	22478	72.8
C. dactylo	Cr		7.68	2.57	0.39	0.0	19630	61.30	12.10	4.22	5560	82693	25267	69.4
Control	Cr		0.0	0.0	0.0	0.0	53804	0.00	0.0	0.0	3400	82683	57201	30.8
T.latiflia	Cd	5812.	1.74	0.31	0.26	0.0	532	90.95	3.75	11.10	768.8	5815	1407	75.8
C. alternif	Cd		5.95	0.72	0.03	0.0	887	35.50	2.14	8.24	666.5	5819	1599	72.5
C. dactylo	Cd		2.61	0.82	0.03	0.0	1005	38.20	2.31	12.03	669.2	5816	1727	70.3
Control	Cd		0.0	0.	0.0	0.0	4139	0.0	0.0	0.0	330.8	5813	4470	23.1
T.latiflia	Cu	9743	1.45	1.00	3.20	0.0	11116	42.90	12.00	39.50	5426.1	97449	16636	82.9
C. alternif	Cu		0.88	1.55	1.82	0.0	16496	24.64	10.12	7.80	4156.9	97460	20695	78.8
C. dactylo	Cu		0.40	2.03	1.70	0.0	19452	16.00	6.60	18.90	5458	97446	24952	74.4
Control	Cu		0.0	0.0	0.0	0.0	51143	0.0	0.0	0.0	3582	97432	54725	43.8
T.latiflia	Zn	77308	1.45	26.95	87.68	0.0	7095	319.92	383.25	316.00	11507	77484	19622	74.7
C. alternif	Zn		0.88	30.14	30.00	0.0	9460	639.22	346.89	167.44	10473	77453	21086	72.8
C. dactylo	Zn		0.40	41.00	42.50	0.0	10287	523.00	273.35	174.20	11269	77438	22527	70.9

Table A1. Continued.

Plant	Heavy metal	Influent, mg	Root (initial) mg	Stem (initial), mg	Leaf (initial), mg	Sediment (initial), mg	Effluent, mg	Root (final), mg	Stem (final), mg	Leaf (f), mg	Sediment (final), mg	Total input	Total output	Removal, %
Control	Zn		0.0	0.0	0.0	0.0	40146	0.0	0.0	0.0	7190	77308	47336	38.8
T.latifolia	Fe	430288	1.45	91.63	103.04	0.0	104533	9831	513.00	1491.0	57792	438861	174159	60.3
C. alternifolius	Fe		0.88	37.41	164.0	0.0	133504	13917	161.84	658.8	63511	437873	211752	51.6
C. dactylon	Fe		0.40	61.65	71.57	0.0	147221	10437	182.60	774.27	47836	43334	206451	52.4
Control	Fe		0.0	0.0	0.0	0.0	309046	0.0	0.0	0.0	29284	430288	338330	21.4
T.latifolia	Pb	271123	1.45	3.31	3.71	0.0	5913	26.66	31.50	25.00	878.64	27132	6874	74.7
C. alternifolius	Pb		0.88	2.88	2.10	0.0	8396	24.80	19.04	10.40	1321.17	27136	9771	64.0
C. dactylon	Pb		0.40	3.60	1.26	0.0	9815	30.0	11.00	6.30	1359.28	27133	11221	58.6
Control	Pb		0.0	0.0	0.0	0.0	20280	0.0	0.0	0.0	291.53	27123	20571	24.2

Table A2. Removal pathway by plants, sediment and other sources

Plants	Heavy metal	Total input, mg	TOTAL OUTPUT, mg	Percent removed by plant %	Percent retained in Sediment %	Contribution by other sources
T. latifolia	Cr	82692.63	15968.41	1.04	40.03	39.6
C. alternifolius	Cr	82701.63	22478.43	0.34	24.61	47.9
C. dactylon	Cr	82693.37	25267.18	0.27	22.00	47.2
Unplanted	Cr	82682.74	57201.38	0.00	5.94	24.9
T. latifolia	Cd	5814.48	1406.73	7.36	54.65	13.8
C. alternifolius	Cd	5818.86	1599.23	2.45	41.67	28.4
C. dactylon	Cd	5815.63	1726.85	2.84	38.75	28.7
Unplanted	Cd	5812.17	4469.49	0.00	7.40	15.7
T. latifolia	Cu	97449.43	16636.02	0.46	32.62	49.8
C. alternifolius	Cu	97459.81	20695.35	0.07	20.09	58.6
C. dactylon	Cu	97445.49	24951.64	0.11	21.87	52.4
Unplanted	Cu	97431.89	54725.08	0.00	6.55	37.3
T. latifolia	Zn	77484.01	19621.47	4.30	58.65	11.7
C. alternifolius	Zn	77452.93	21086.23	4.78	49.67	18.3
C. dactylon	Zn	77437.86	22527.24	3.73	50.02	17.2
Unplanted	Zn	77308.05	47335.99	0.00	15.19	23.6
T. latifolia	Fe	438860.48	174159.07	1.87	33.18	25.3
C. alternifolius	Fe	437873.20	211751.81	3.38	29.99	18.3
C. dactylon	Fe	433339.73	206451.19	4.04	23.17	25.1
Unplanted	Fe	430287.71	338329.62	0.00	8.66	12.7
T. latifolia	Pb	27131.91	6874.29	1.09	12.78	60.8
C. alternifolius	Pb	27136.25	9771.16	0.42	13.52	50.0
C. dactylon	Pb	27133.34	11221.32	0.33	12.11	46.2
Unplanted	Pb	27123.44	20571.39	0.00	1.42	22.7

List of publications

Peer-reviewed Journal Papers

- **Mustapha, Hassana Ibrahim.**, van Bruggen, J. J., Lens, P. N. (2015). Vertical subsurface flow constructed wetlands for polishing secondary Kaduna refinery wastewater in Nigeria. Ecological Engineering. 84, 588-595. doi:10.1016/j.ecoleng.2015.09.060

- **Mustapha, Hassana Ibrahim.**, van Bruggen J.J.A., Lens P.N.L. (2018), "Fate of heavy metals in vertical subsurface flow constructed wetlands treating secondary treated petroleum refinery wastewater in Kaduna, Nigeria", International Journal of Phytoremediation, 20(1): 44-53. doi: 10.1080/15226514.2017.1337062

- **Mustapha, Hassana Ibrahim.**, van Bruggen J.J.A., Lens P.N.L. (2018), "Optimization of petroleum refinery wastewater treatment by vertical flow constructed wetlands under tropical conditions: plant species selection and polishing by a horizontal flow constructed wetlands" Water, Air and Soil Pollution. 229(4):137-154. https://doi.org/10.1007/s11270-018-3776-3.

- **Mustapha, Hassana Ibrahim.**, Gupta, Pankaj Kumar., Yadav, Brijesh Kumar., van Bruggen J.J.A., Lens P.N.L (2018), "Performance evaluation of duplex constructed wetlands for the treatment of diesel contaminated wastewater". Chemosphere. DOI:10.1016/j.chemosphere.2018.04.036.

- **Mustapha, Hassana Ibrahim.**, van Bruggen J.J.A., Lens P.N.L. (2018), Vertical subsurface flow constructed wetlands for the removal of petroleum contaminants from secondary refinery effluent at the Kaduna refining and petrochemical company (Kaduna, Nigeria). Accepted by the Journal of Environmental Science and Pollution Research for publication.

- **Mustapha, Hassana Ibrahim.**, Lens P.N.L. (2018), "Constructed wetland systems for treatment of petroleum refining industry wastewater: A review" Submitted to a Journal for possible evaluation.

Conference Proceeding

- **Hassana Ibrahim Mustapha**, Rousseau, D., van Bruggen, J., & Lens, P. (2011). Treatment performance of horizontal subsurface flow constructed wetlands treating inorganic pollutants in simulated refinery effluent. *In:* Book of Proceedings of the 2nd Biennial Engineering Conference of School of Engineering and Engineering Technology, Federal University of Technology, Minna, held in November 16th – 18th, 2011 at Federal University of Technology, Minna, Nigeria.

- **Hassana Ibrahim Mustapha**., Bruggen van J.J.A., P. N. L. Lens., 2013. Preminary studies on the application of constructed wetlands for a treatment of refinery effluent in Nigeria: A mesocosm scale study. In: Book of Proceedings of the *2013 International Engineering Conference, Exhibition and Annual General Meeting* of the Nigerian Society of Engineers' held in December 9th - 13th, 2013 at the International Conference Centre, Abuja. Nigeria. Abuja: Nigerian Society of Engineers.

- **Hassana Ibrahim Mustapha**., Bruggen van J. J. A., Lens P. N. L. (2013). Treatment of Refinery Effluent Using Vertical Subsurface Flow Constructed Wetlands: A Case Study in the Tropics (Book of Abstract, O.144). Oral presentation at the 5th International Symposium on Wetland Pollutant Dynamics and Control, WETPOL2013, Nantes, France. October 13 – 17 2013.

- **Hassana Ibrahim Mustapha** (2016). Invasion, management and alternative uses of *Typha latifolia*: A brief review. In: Book of Proceedings of the Nigeria Institute of Agricultural Engineering Conference, Minna. Held in October 2016 at the Federal University of Technology, Minna, Nigeria.

- **Hassana Ibrahim Mustapha** and P. N. L. Lens., 2017. Treatment of Secondary Refinery Wastewater by *Cynodon dactylon* planted Horizontal Subsurface Flow Constructed Wetland Systems. In: Book of Proceedings of the Association of Professional Women Engineers of Nigeria (APWEN) International Conference held in September 25th – 28th, 2017 at Kano, Nigeria.

- **Hassana Ibrahim Mustapha** and P. N. L. Lens., 2017. Polishing of secondary refinery wastewater for reuse purposes. In: Book of Proceedings of the 2nd International Engineering Conference (IEC 2017), Federal University of Technology, Minna, held in November 16th – 18th, 2017 at Federal University of Technology, Minna, Nigeria.

About the author

Hassana Ibrahim Mustapha was born on the 12th of November 1974 in Lokoja, Nigeria. She is an Agricultural Engineer by training. She obtained her B.Eng. in Agricultural Engineering and M. Eng. (Soil and Water Conservation Engineering) degrees from the Federal University of Technology (FUT), Minna, Nigeria in 2000 and 2005, respectively. Since her Master's graduation (2005), she works for Federal University of Technology (FUT), Minna (to date).

Hassana Ibrahim Mustapha joined the UNESCO-IHE, Institute for Water Education in the Netherlands in 2011 as a PhD Research Fellow. Her research work focused on the treatment of petroleum refining wastewater. Her PhD was supervised by Professor dr. ir. Piet N. L. Lens (Professor of Environmental Biotechnology). She has co-authored 12 papers on constructed wetlands for petroleum refining wastewater treatment, including 6 conference papers. Hassana's research interests are in constructed wetlands, phytoremediation, water quality and treatment of wastewaters.

Acknowledgements of financial support

I wish to acknowledge the Netherlands Government for providing the funding through NUFFIC (NFP-PhD CF7447/2011) for this research, this is the genesis of the PhD degree. I acknowledge also TEDFUND for their bench work support through my institution, Federal University of Technology, Minna. Nigeria.

Netherlands Research School for the
Socio-Economic and Natural Sciences of the Environment

D I P L O M A

For specialised PhD training

The Netherlands Research School for the
Socio-Economic and Natural Sciences of the Environment
(SENSE) declares that

Hassana Ibrahim Mustapha

born on 12 November 1974 in Lokoja, Nigeria

has successfully fulfilled all requirements of the
Educational Programme of SENSE.

Delft, 29 June 2018

the Chairman of the SENSE board

Prof. dr. Huub Rijnaarts

the SENSE Director of Education

Dr. Ad van Dommelen

The SENSE Research School has been accredited by the Royal Netherlands Academy of Arts and Sciences (KNAW)

K O N I N K L I J K E   N E D E R L A N D S E

The SENSE Research School declares that **Hassana Ibrahim Mustapha** has successfully fulfilled all requirements of the Educational PhD Programme of SENSE with a work load of 39.9, including the following activities:

SENSE PhD Courses

o   Environmental research in context (2011)
o   Research in context activity: 'A seminar presentation on the potentials of constructed wetlands for industrial wastewater treatment organised in Minna, Nigeria' (2014)

Other PhD and Advanced MSc Courses

o   Nanotechnology for water and wastewater treatment, UNESCO-IHE Delft (2011)
o   Wetlands for wastewater treatment, UNESCO-IHE Delft (2011)
o   How to Write a World-Class Paper, Wageningen University (2011)
o   Award winning research grant proposal writing workshop, Federal University of Technology in Minna, Nigeria (2016)
o   Workshop on academic and technical writing, Federal University of Technology in Minna, Nigeria (2016)

External training at a foreign research institute

o   Test and measuring organic and inorganic parameters by gas chromatography and spectrophotometer, Federal University of Technology in Minna, Nigeria (2016)

Management and Didactic Skills Training

o   Supervising of three MSc students Federal University of Technology, Minna, Nigeria (2012)

Selection of Oral Presentations

o   *Treatment of refinery effluent using vertical subsurface flow constructed wetlands: a case study in the tropics.* 5th International Symposium on Wetland Pollutant Dynamics and Control, WETPOL, 13-17 October 2013, Nantes, France
o   *Preliminary studies on the application of constructed wetlands for treatment of refinery effluent in Nigeria: A mesocosm scale.* Name of Conference, Nigerian Society of Engineers' International Engineering Conference, Exhibition and Annual General Meeting, 9-13 December 2013, Abuja, Nigeria
o   *Treatment of secondary refinery wastewater by Cynodon dactylon planted horizontal subsurface flow constructed wetland systems.* 2017 International Conference and annual general meeting of the association of Professional Women Engineers of Nigeria, 26-28 September 2017, Minna, Nigeria

SENSE Coordinator PhD Education

Dr. Peter Vermeulen

Printed and bound by CPI Group (UK) Ltd, Croydon, CR0 4YY

21/10/2024

01777112-0016